2014 AMSOIL/*STREET RODDER* ROAD TOUR SPONSORS

AMSOIL STREET RODDER 2014 ROAD TOUR

PRESENTED BY ARP automotive Racing products

ARP automotive Racing products

AEROMOTIVE

AMSOIL

COMP CAMS

Classic INDUSTRIES Parts & Accessories

custom autosound MANUFACTURING, INC.

Dakota Digital

1-800-633-5559 DYNOTECH DRIVESHAFTS Dynotecheng.com

www.fuelairspark.com FAST FUEL AIR SPARK TECHNOLOGY

FLAMING RIVER Quality Steering Systems

HOTROD INTERIORS by GLENN KRAMER 623-872-2242

HotRods BY DEAN

CIRCLE RACING WHEELS

Xtreme Dynamat

Eastwood DO THE JOB RIGHT.

FLOWMASTER THE EXHAUST TECHNOLOGY COMPANY

HUGHES PERFORMANCE

Intercity Lines, Inc

JET-HOT HIGH PERFORMANCE COATINGS

Miller

PPG

PAINLESS PERFORMANCE PRODUCTS

RS ROADSTER SHOP

RS ROADSTER SHOP BY PENSKE

SHERM'S CUSTOM PLATING SACRAMENTO WHEEL

Scat Crankshafts

Shaver Specialties RACING ENGINES

SOFF SEAL Weatherstripping www.soffseal.com

SPW

SINCE 1952 SPEEDWAY MOTORS America's Oldest Speed Shop

U.S. Radiator

wilwood DISC BRAKES

LOKAR PERFORMANCE PRODUCTS QUALITY... PLAIN & SIMPLE

PROCAR

VINTAGE AIR

STREET RODDER MAGAZINE

GUIDE TO BUILDING A '59 CHEVY

Hot Rod Network
Street Rod Group

Street Rod Group Director
Brian Brennan

Senior Tech Editor
Ron Ceridono

Written By
Tim Bernsau, Brian Brennan, Gerry Burger, Ron Ceridono, James Humble, and Jeff Smith

Photography By
Tim Bernsau, Brian Brennan, Randy Lorentzen, and Robert McGaffin

Designers
Martin McHugh, Chris Bywater

Copy Editor
Sarah Gonzales

General Manager
Tim Foss

Associate General Manager
Janeen Kirby

Executive Assistant/ Event Coordinator
Yasmin Fajatin

Design Director
Markas Platt

Road Tour Director
Jerry Dixey

ISBN: 978-1-935231-32-5

NOTICE: The information in this book is true and complete to the best of our knowledge. All recommendations on parts and procedures are made without any guarantees on the part of the author or the publisher. Tampering with, altering, modifying, or removing any emissions control device is a violation of federal law. Author and publisher disclaim all liability inured in connection with the use of this information.

Contents

CHAPTER ONE

Hot Rods By Dean

Builders of Our '59 Chevy Impala

By Ron Ceridono | Photography by Brian Brennan

One of the fastest growing names in the street rod industry is Dean Livermore of Hot Rods by Dean. Located in Phoenix, Livermore and his crew of 11 craftsmen occupy a 15,000-square-foot shop where they do everything from basic street rod maintenance to ground-up construction of turnkey cars. Hot Rods by Dean built our 2005 Road Tour car and we were lucky enough to have the crew back again to do our '59 Chevy for 2014.

Livermore's interest in all things mechanical began early and can be traced to a broad range of experiences—his dad was in the aerospace industry and his uncle owned a service station and Livermore was working with both of them by age 9. After high school Livermore attended Universal Technical Institute where he went through the auto/diesel/transmission programs, which led to working for CAT and later as a dealership line mechanic. While earning his living twirling wrenches, Livermore was building his own cars at home and was surprised to find that some people would actually pay to have street rods built. That realization led to a seven-year stint working in a street rod shop and that was where Livermore met one of the pioneers of the hobby, Dick "Magoo" Megugorac.

Ironically Livermore didn't know who Magoo was when he volunteered to wire Megugorac's Merc after work in exchange for a few dinners. While hanging out together it didn't take long for them to recognize each other's talents and the two became fast friends. With Magoo as his mentor, Livermore opened the doors to Hot Rods by Dean in 2001 and he was on the way. By 2005 Livermore had won the coveted Trendsetter Award, presented at the Hot Rod Industry Alliance reception at the SEMA show. Since then Livermore has gone on to build the World's Most Beautiful Custom at the 2007 Sacramento Autorama, took a '67 GMC pickup to compete in the Radical Pickup class at the Detroit Autorama, and found it running for the Ridler Award by making the Pirelli Great 8.

Livermore has gone on to build a long list of award winners of every description—street rods, customs, trucks, Cobra kit cars, and sports cars, just about anything on wheels. And while the type of vehicle rolling out the shop doors varies, they all exhibit the same quality workmanship and attention to detail HRBD has become known for. Certainly that was a significant factor in choosing Livermore and company to build the latest Road Tour car; we needed someone who could combine beauty and brawn. With the Road Tour crisscrossing the country, our Chevy will accumulate approximately 25,000 miles in one summer, which means the car has to be comfortable when subjected to weather extremes and lousy road conditions. It has to be reliable and capable of gobbling up mass quantities of asphalt in a single session while still looking great. In short, the car has to be perfect in all respects, it has to be part prom queen and part running back.

We know that creating something like the AMSOIL/*STREET RODDER* Road Tour Chevy takes talent and commitment. Given the short time frame available and the number of parts suppliers and sponsors involved, there are constant scheduling and organizational challenges to deal with, not to mention the "last minute" decisions that have to be made during construction. Despite the challenges, Dean Livermore and the team at Hot Rods by Dean not only met our expectations, they exceeded them and gave us a car we're truly proud of. Well done.★

❯ Thanks to the incredible talents of the Hot Rods by Dean staff and a long list of quality aftermarket parts suppliers the 2014 AMSOIL/*STREET RODDER* Chevy is ready for the 19th edition of the Road Tour.

HotRods
BY DEAN

CHAPTER TWO

Development

How GM's Radical '59 Chevrolet Came To Be

By James "Wick" Humble

Even as the epochal and precedent-breaking '55 Chevy line was causing sensations at the dealers' showrooms and racetracks, GM management had already decreed the first step in a market strategy to upsize and upgrade all its products. Well, let's face it: in the '50s the automotive marque to which the overwhelming majority of the automotive masses aspired was the General's flagship: Cadillac. Traditional rival Packard was going into its final coma, Lincoln was only a spunky wannabe, and Chrysler's Imperial was a distant "me, too." The Caddie, in the eyes of the public, held the whip hand in terms of luxury, speed, and "charisma"—a quality that is hard to define, but sells cars. The tail wagged the dog and the "Low-Priced Three" were dog itself.

These were the days when new-car introduction time was like a circus in town. When its arch competitor Ford and upstart Plymouth both came out with all-new bodies and frames in 1957, ever wonder why big Chevrolet didn't follow suit? It's too easy to assume that as king of the Low-Priced Three market, Chevy merely wanted to

prolong the line of that classic with the iconic '57; nope—the front office had decreed that Chevrolet would spring a big (literally) surprise on the market to claim lasting precedence in 1958! Ford's all-new—if traditional—body thus sold very well, and controversy exists to this day who won the 1957 sales race; suffice it to say that Chevrolet not only didn't like to lose, they likewise didn't relish a close finish. This surprise was a new line-leader called the Impala, and it would bring even more Caddie cachet to the arena. It was to be a package: a completely new, compact V-8 at 348 ci (available with dealer-installed Borg-Warner four-speed), a new X-frame that moved the engine's harmonic balancer up even with the spindle centerline—all in support of a more luxury-oriented, and appreciable heavier, body. Released only as a convertible and two-door hardtop—remember that the '58 Bel Air series had its own hardtop that shared few body panels—it overwhelmed the warmed-over second-year styling of both Ford and Plymouth, and notched up the luxury target for years. And, in a hotly contested race for high sales, it regained the winner's circle—but the challenge would be how to follow that act?

Precedent was broken a year later when GM planners pre-empted the new bodies (and similar Pontiacs as well) with a radically new design that retained only the basic frame and power packages of the '58s. Why? Well, part of the plan was to get all the divisions in synch over at Fisher Body with intensified inter-line rationalization: think roof and window shapes. After that, C-B-O-P-C could march forward in a unified rank, thus significantly simplifying parts inventories and saving the bean counters lots of legumes. At least that was the theory. To keep these developments in context, recall that a wild and hyper-controversial new

Full-sized Clay 'mode
Would have had on rear bumper with six round lites enclosed.
"Flying Vee"
what a huge trunk
Credit GM Photograph
date: 12-12-'56
for 1959

Early Clay; and we thought the Edsel was WILD!
~~Dash~~ Grille texture went on 1961 Pontiac - center head lites in trash.
Credit GM Photograph
date: 7-5-'57

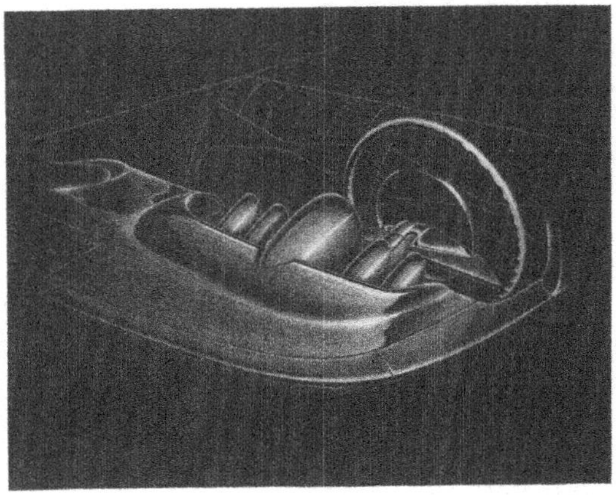

White Prismacolor or crayon on black illustration board:
That radical dash evolves!
Credit "GM Photograph
Undated—
for 1959

Rear view fiberglas prototype 'model' 1:1 scale. "Cats-Eye" tail lites, anyone?

Credit: GM Photographic
date: 10-25-57
for 1959

Not a real car - fiberg, model but full-sized Corvette-style grille lives! "Factory Approved"

Credit GM Photographic

date: 10-25-57
for 1959

"Blue Sky" rendering from 1957: double grille openings, covered front wheel-wells, impossibly elaborate rear fender treatment shows how limits were off for '59's where are these paintings now?!

lineup was being readied by FoMoCo in 1958; that styling boomerang called the Edsel! Sensation proved its worth, and the E-Car sold very well at first, but mostly to the detriment of Ford and Mercury. Few now recall the poor timing of it all; 1958 was the year of the "Eisenhower Recession" and though the new Chevy again prevailed in its market, overall car sales slumped badly, and adventurous marketing proved hard sledding!

For 1959, GM Design Center would set the "look" and basic body parameters for a number of wheelbases, and engineering could fix all the crucial body dimensions—firewall and door hinge lines, floor and roof height, and so on—and the five division studios could take it from there. Looking back on the photo files of GM Photographic a marvelous nuthouse of ideas can be revealed, with lots of themes and treatments recognized that later crossed marque lines. Of course, the individual divisions had always expected to adapt to some degree of rationalization and corporate identity, but in those dear, dead days, in-house politics and sales competition being what they are, each nameplate still fought the trend like the Army fights the Navy fights the Air Force!

With full management approval, Chevrolet stylists took a huge gamble that those huge horizontal fins, even in a tailfin crazy market, would catch the fancy of the new-car buyer. Wide "cat's-eye" taillights never prototyped on any GM dream car drove home the impression. Controversy didn't hurt, in the real-time market, as all sorts of rumors surfaced about the '59 cars becoming airborne (or at least unloading the rear springs) in high-wind areas: supposedly a coastal highway tunnel had the capability to cause the lifting of the rear suspension to a really thrilling

front end rather sim. to '60 Buick, but open front wheel wells diff. from other airbru rendering. Narr w/w's didn't appea until 1961 on Chev Just needs angel-h

date 1957

extent. Aerodynamicists scoffed, noting that the airfoil curve that causes lift was totally lacking, but it became an urban legend. As the dropped-in-front "Dagoed" look was beginning to move from street rods to late-models, the sight of radically raked '59s might have nurtured this rumor.

Frontal styling for the new lineup was a bit controversial for a different reason: critically viewed, it didn't quite "match" the rear look as seen in all previous Chevys. Though pleasing in a less radical way, to some critics it looked as if it had been chosen "off the rack" by the stylists and merely attached to the distinctive tail by the expedient of slathering clay to join two design bucks. Some definite Cad influence can be discerned in the stamped-aluminum grille texture; a toothy 'Vette-inspired unit was considered! The upside was that the treatment lent itself to loads of effective restyling, which customizers, even the high school auto shop kind, loved. The most popular of these was the replacement of the designers' nice "grin" with a simple tube grille of chromed bars. As to the twin pseudo-air scoops at the hood parting line, one custom shop artist [I think it was Darryl Starbird]

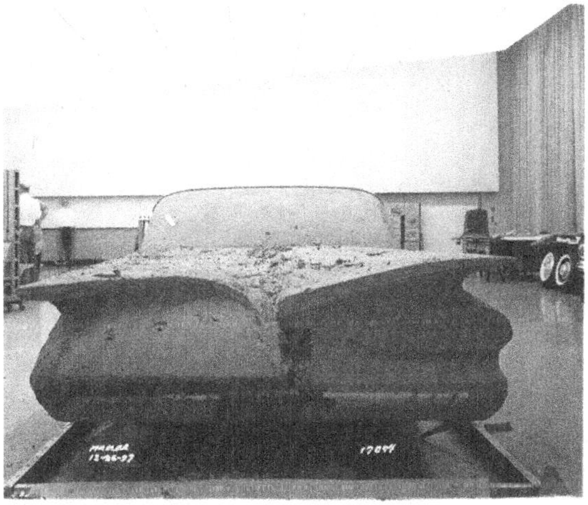

Two treatments to evolve the final '59 design — right side very close, bu left not bad, ear hint of '60 directio Note scraps; clay modelers must ha been wizards!

date: 12-6-57

Wait: wasn't this style from the REAR fenders of 1958 Impala? Package 'bucks' were reused — cl scraped off and new idea modele Prob. same as 12/19

date: 1-3-58

Clean design; fron
and rear integrat
Det. 'bubble-top' ro
established for cou
'B-47 jet pod' influ
popular in Buick
Studio – seen on
'59-60. Kustom boy
would have enjoye
this metal!
date: 12-19-57

Multi-fin design
from rear. Ponti
got 1959 Bonnevill
tail lites from it,
possibly.
Horrible proportio
fortunate this cla
got put back in t
bin!
date: 4-9-57

"Wing" fin theme
evolving on 1:1 clay –
chrome-foil being
applied to edges.
Deep, heavy bump
fights deck treatm
"Detroit – as drawn
by MAD magazine."
date: 7-17-57

THE '59 COUPE ROOF WAS PRETTY AND COMPLEMENTED THE LOOK PERFECTLY.

remarked, in noting how little he cared for the effect, that it looked like one of the body designers had stood too close when holding a running grinder, and having buzzed off the leading edge, had to repeat on the other side for the sake of symmetry!

The '59 coupe roof was pretty and complemented the look perfectly; it still looks sweet, if you grant that the rear seat occupants might get a bit warm. Thought the "Forward Look" Chrysler products had sported a modest bubbletop look as early as 1957, but GM stylists trumped it with a super-thin C-pillar and beach-towel sized roof panel on all their two-door hardtops, ending it with the low-production and even prettier Bel Air Sports Coupe in 1962. Even the two-door sedan roofs looked rakish. All divisions had an optional four-door hardtop profile that was convenient, but rather bland. While the fin design was no particular help at the new Daytona oval, the roof profile was a real plus; unfortunately, Pontiac and Olds had it too. With a well-worn '59 Sports Coupe "Junior" Johnson did win the 1960 500, but it had been discovered that the "W" head/combustion chamber design became a bit breathless on the high-speed oval. Fortunately, it proved to be a torquer that always did better on short tracks or drags.

Performance-wise, neither year had it all their way in the Stock Car wars; there was plenty of competition, especially from GM rival Pontiac with a larger displacement 389ci motor that breathed very well and rode in a perimeter frame with a wider stance. The second-gen four-seat Thunderbird was low, had a smaller frontal area, and was equipped with the big Lincoln 430ci powerplant. However, at the now burgeoning dragstrips, the Chevrolets did very well, and seemed to be everywhere garnering NHRA class records, both with the 348 and the 283 ci in lower echelons. That the cars were campaigned hard and in great numbers is told by a perusal of the strip box scores from your old *National Dragster* weekly. Bear in mind that despite the fact that the "W" was actually a tad lower than the small-block, it was more massive and couldn't really be shoehorned into the elderly frame of GM Managements' performance car of choice, the sporty Corvette.

From its introduction, the W benefitted strongly, as did many GM cars, of the availability of the Borg-Warner T10 four-speed (originally developed from their heavy-duty three-speeder), especially in the upper stock/stick classes at the local strip. Quick: what carburetion equipment (not counting FI) was never offered on the "W-motor"? That the engine

Another Niel Madler (GM Photographic) shot of ugly duckl. '59 proposal. Chevy fans dodged a big bullet on the design!

date: 4-9-57

Early clay 1:1 model '58 design was set by now; so where did this awkward proposal evolve!? Homely as a 1960 Dodge!

date: 4-9-57

Small-scale clay; Very early proposal and rear fender theme is alive. Note: two differen[t] roof/windslield treatments on one model.

date: 1-11-57

Early clay – very conservative. Infl. 1959-60 Olds? Modeler still working clay on rt. front bumper area. Could have been worse!

date 1-11-59

Front-end design fairly set by now 'jet-pod' headlite treatment found on Buick later. Rendering on wall shows slightly diff interpretation.

date: 1-30-58

There is '60 design emerging, but also hints of '61 rear end as well. ← What is under drop cloth in rear of shot? Chrysler influence, if not actual MoPar!

date: 2-20-58

was never sold with a basic two-barrel carburetor bolsters the stance propounded in the prestigious Society of Automotive Engineers "Engineering Journal" of 1957 that Chevy boffins were on the hunt for more BMP and torque with the 348 series! Not many V-8s can make that claim; the "W" came only with the basic Rochester 4GC four-barrel, or optional triple Rochester 2GC two-barrel intakes. Later, of course, the 3x2s were supplanted by two hard-to-quench Carter AFB four-barrels, but that was with the rather tardy advent of the famed 409ci iteration. Why did the "W" get the appellation of "truck motor"? Like the original 265ci mouse-motor, it was introduced in both automobiles and commercials simultaneously; but the 409 was retained in Chevy's larger gas-powered trucks after it had been supplanted by the 396-427ci big-block in passenger cars. Chevrolet engineers stated in the *SAE Journal* that based on the metallurgy available in 1955-56, they didn't think the small-block could be reliably built beyond 301 ci. But recall that the competition's little V-8 never went beyond 312 ci.

Whereas division rival Pontiac held its own against the rapidly developing threat of the B-block 413ci Chrysler products vary, especially with automatic-equipped cars becoming heavy threats for top stocker after the advent of the Torqueflite, Chevy had really needed the 409 at least a year earlier. And the ancient two-speed Powerglide, never a threat in stock configuration, had to soldier on in the automatic classes for too long. The 1959-60 Chevrolet, especially the Impala Sports Coupe, was the ideal car to cruise main and attract attention; under the streetlights, it screamed (especially in Gypsy Red lacquer, strong forward rake, and playing a tune on dual Pacemaker mufflers and the three-twos intake) "Hey! Look at me!" With all the windows rolled down, of course, cat's-eyes glowing to the rear, and AM radio thumping out a monaural "Little Darlin'" or Jorgen Ingmans' "Apache," one was automatically trolling for black and whites!

The new body essentially lasted two years; in 1960 the boys in the styling backroom had toned down the flamboyant fins a bit, making the decklid a bit flatter, and reinstated the modest round taillights that had arrived two years earlier, and these soon became a brand identifier even on Corvette and Corvair. Also, they decreed a pleasing new, less radical grille shape that was tied quite neatly into the rear treatment with a high horizontal reveal line. Otherwise, the '60 lineup was almost identical to the '59 in most respects, and despite the pretty new Ford styling, featuring gullwing horizontal fins and bubble-roof, interestingly it took the sales trophy. Before summing up, we should add that the Motorama-inspired dash/instrument treatment (besides the inclusion of "idiot lights") was a grand slam of a design!

Great illustration of evolution from '5 to '60 fins! I prefer this treatment to what ca out on both! Rakish, subtle (sort appealing.

date: 1-23-58

The wild-styling card was discarded by 1961. That marked the advent of non-controversial body styling—starting a generation of Chevrolet domination in sales of what had become known as fullsized family cars. And a half-year intro of the balladized 409 (and rare Z-11 427) the 348 W was progressively supplanted by the legendary 327ci SBC, and the awesome Rat motor—produced partially on the same assembly lines as the W—was just around the corner. Well, heck: even the Caddies had become conservative and trimmed of fin by then. But now, as then, the '59 Chevy turns heads as a street cruiser on Saturday nights! *

1960 Coupe shape def. set by mid-58

Didn't that grille treatment resurface on early Chevelle S.S?

Note how stylists finally pulled front and rear designs together at last.

date 5-19-58

CHAPTER THREE

X-Frame Exorcism

A New Roadster Shop Chassis and Suspension

By Ron Ceridono | Photography by Robert McGaffin

From the moment '59 Chevys hit showroom floors there was no mistaking them for anything else. Everything about their appearance was unique, particularly when viewed from the rear. We've always had a fondness for these cars, and as we couldn't think of anything cooler to drive on the AMSOIL/STREET RODDER Road Tour. We found a solid specimen and turned it over to Dean Livermore at Hot Rods by Dean to get it Road Tour ready.

While we admire the looks of our '59, in this case beauty truly is only skin deep—underneath the stylish sheetmetal is a truly hideous chassis. First found under '57 Cadillacs the X or cruciform frame was also found under '58-64 Chevys. These frames had an hourglass shape and GM boasted that the design allowed the seating area to be lowered, which in turn made for a lower roofline. Although there was some truth to that claim, one of the lesser-known in-house arguments for the X-shape was cost effectiveness. The same basic frame could be used under a variety of cars, as simply changing the length of its centersection easily altered the wheelbase.

From a functional standpoint these frames had some significant shortcomings, not the least of which

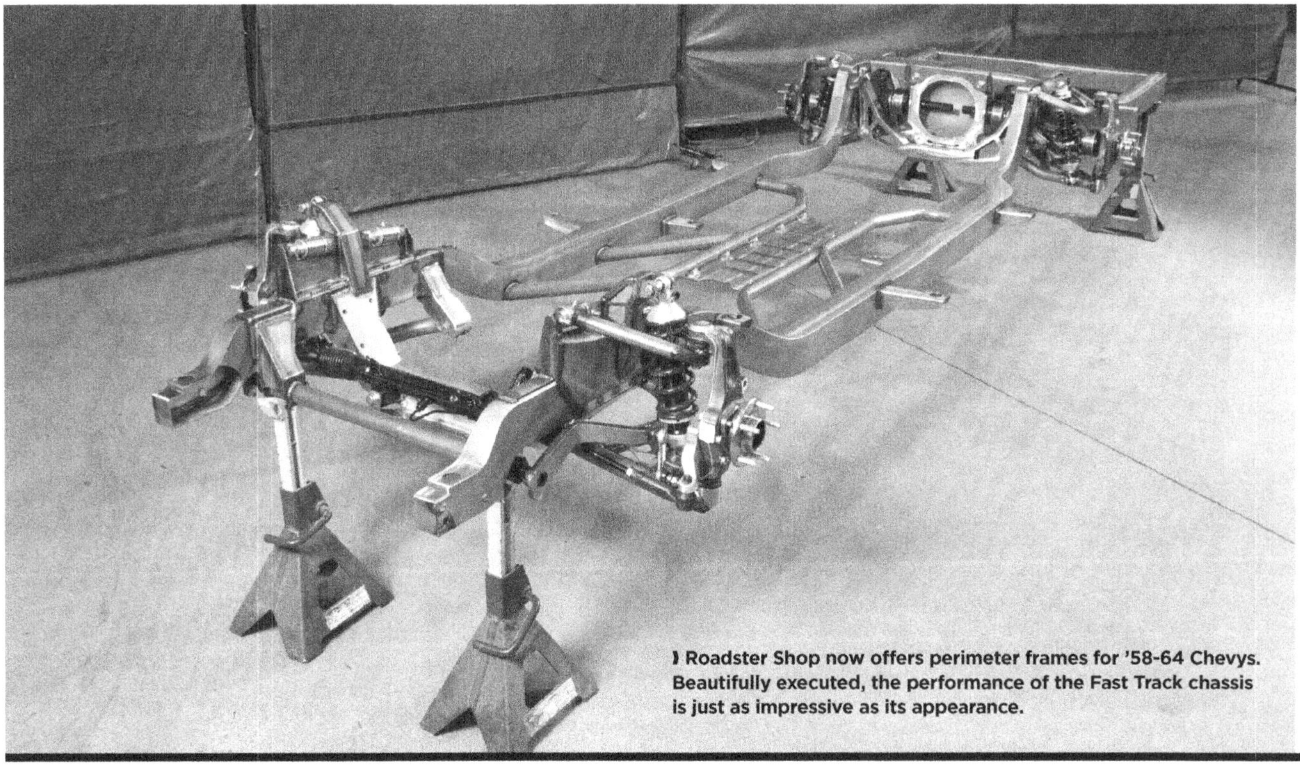

Roadster Shop now offers perimeter frames for '58-64 Chevys. Beautifully executed, the performance of the Fast Track chassis is just as impressive as its appearance.

This is the original '59 Chevy cruciform, or X-frame. Why it lacks resistance to twist and offers little protection in the event of side impact is easy to see.

Using a laser the full-length pieces of the framerails are cut from a sheet of 10-gauge cold rolled steel.

Zach Flores begins welding the 'rails together. By fabricating the frame out of separate pieces the complicated contours required to fit a perimeter frame under '58-64 Chevys without extensive floor modifications can be created.

All the fabrication is done in a precision fixture. Here the inside boxing plate is being added to a framerail.

The stock body mount locations are used; additional mounts that align with reinforced sections of the body have been added as well.

is the lack of torsional rigidity. While a floppy frame wasn't a big deal with the squishy suspension these cars came equipped with, to accommodate the high-performance underpinnings a much more solid foundation would be required. We got the high-performance IFS and IRS we wanted and the new frame to hang them on from Roadster Shop.

The foundation for our Road Tour car is one of Roadster Shop's Fast Track perimeter frames, which they offer for '58-64 Chevys. Each framerail is fabricated from four separate pieces of laser-cut 10-gauge cold-rolled steel welded together in a fixture. The completed frames are far stronger than the original design and require very minimal floorpan modifications to fit.

On each end of our new frame, Roadster Shop's Fast Track independent suspension will be found. As handling is a priority, allowances have been made to fit lots of rubber under the stock sheetmetal at all corners; there's room for 10-inch wheels with 285 tires up front and 12-inch wheels with 345 tires in the rear.

Rather than just graft on an existing IRS, the Fast Track rear suspension was engineered by Roadster Shop for cars of this size and weight while complementing the

6

During the assembly procedure dimensions are checked and rechecked to ensure accuracy.

7

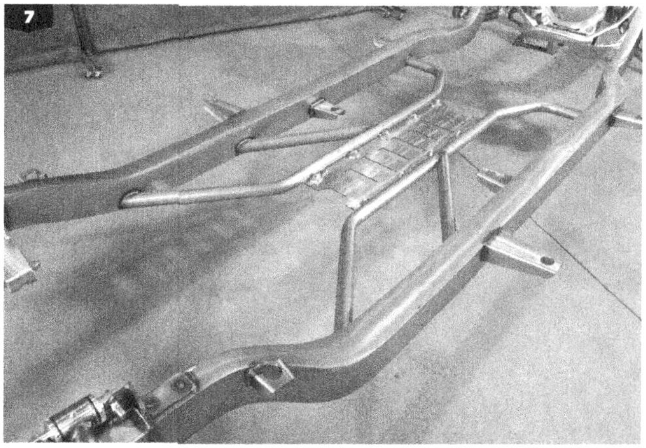

With the 'rails fabricated and the body mounts in place, the center crossmember is installed. The perimeter frame offers more room for the engine, transmission, and exhaust as well as eliminating the need for a two-piece driveshaft.

8

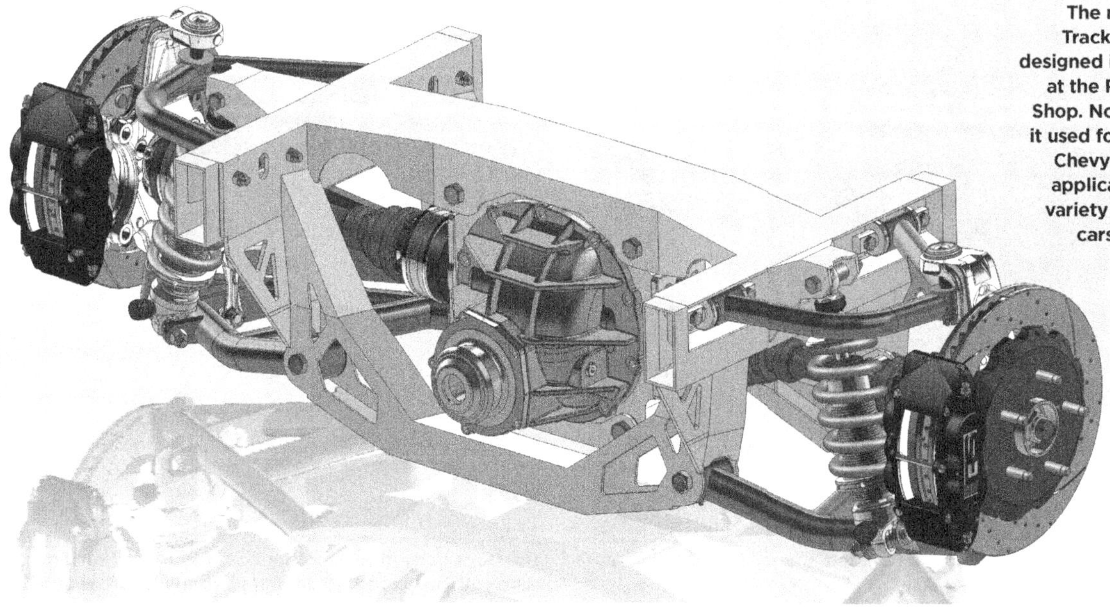

The new Fast Track IRS was designed in-house at the Roadster Shop. Not only is it used for '58-64 Chevys but it's applicable to a variety of other cars as well.

tread width of the Fast Track IFS. The centersection is based on the venerable 9-inch Ford, heavy-duty CV joints, driveshafts are used, and the spindles are C6 Corvette.

Phil Gerber, VP of marketing for the Roadster Shop, explains, "Penske has long been the benchmark for high-performance shock absorbers. Instead of using a product that compares themselves to the best, we wanted to use the best in an effort to continue to bring our customers the highest quality and highest performance product out there."

For a solution to the original front suspension system's shortcomings, we chose the Roadster Shop's new Fast Track IFS.

Based on C6 Corvette spindles and Z06 hub assemblies, Roadster Shop has optimized the suspension geometry for applications like our Chevy. Handling has been enhanced by increasing camber gain during hard cornering and lowering the roll center. Bumpsteer through the full range of suspension travel is practically immeasurable and to control body roll there is a 1.25-inch-diameter splined antiroll bar with Corvette endlinks to keep the chassis flat when negotiating corners.

Standard on all Fast Track suspension systems and complete chassis are non-adjustable, RS by

9 A cradle system is used to mount the Fast Track IRS. Here it is being fitted to framerails.

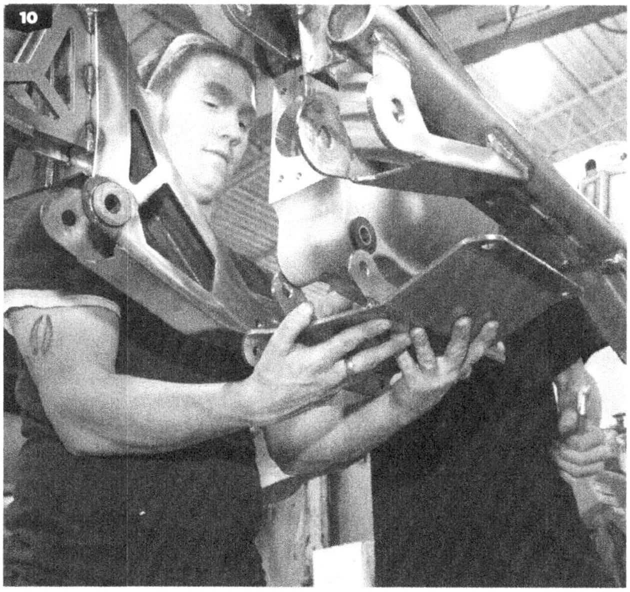

10 With the cradle in position assembly of the suspension can begin. Here the 9-inch housing is bolted in place.

11 As with the front suspension, upper and lower control arms are used (here the uppers are being put in place).

12 These are the stub axles and the retainer plates that hold them in the centersection.

Penske shocks that are custom valved for the best possible ride quality and handling characteristics. Upgrades available include double-adjustable shocks as well as double- and triple-adjustable remote reservoir shocks for those who are looking for the utmost in tunability and performance.

Basically the RS Fast Track IFS is everything the original wasn't. It provides the handling characteristics we wanted with an improved ride and a much lower stance while still providing full suspension travel. The RS Fast Track IFS is available with a universal crossmember designed for frames with 29 inches between the 'rails and comes with track widths of 54.75 or 58.75 inches. It's also available on the '58-64 Chevy replacement frames, as shown here. ★

The stub axles are secured to the aluminum center housing by six bolts. The CV joints attach to the flanges on the stub axles.

Half shafts are equipped with heavy-duty 930 and 934 constant velocity joints. Shafts are available to handle up to 1,400 hp.

The spindles, hubs, and ball joints are C6 Corvette; Roadster Shop manufactures the control arms.

All Fast Track chassis come standard with non-adjustable RS by Penske shocks that are valved for the specific application. We opted for the double-adjustable versions.

All the components of the Fast Track IRS, including the coilovers, are easily accessible for adjustment or maintenance.

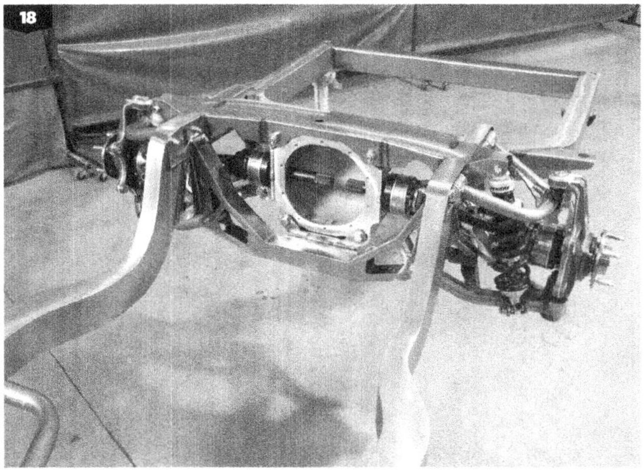

The complete Fast Track IRS without brakes. Wilwood brakes for C6 Corvettes are a direct fit.

Rear camber adjustments are made at the upper A-frame attachment points. Toe corrections are made with the hex-shaped link that connects to the upright.

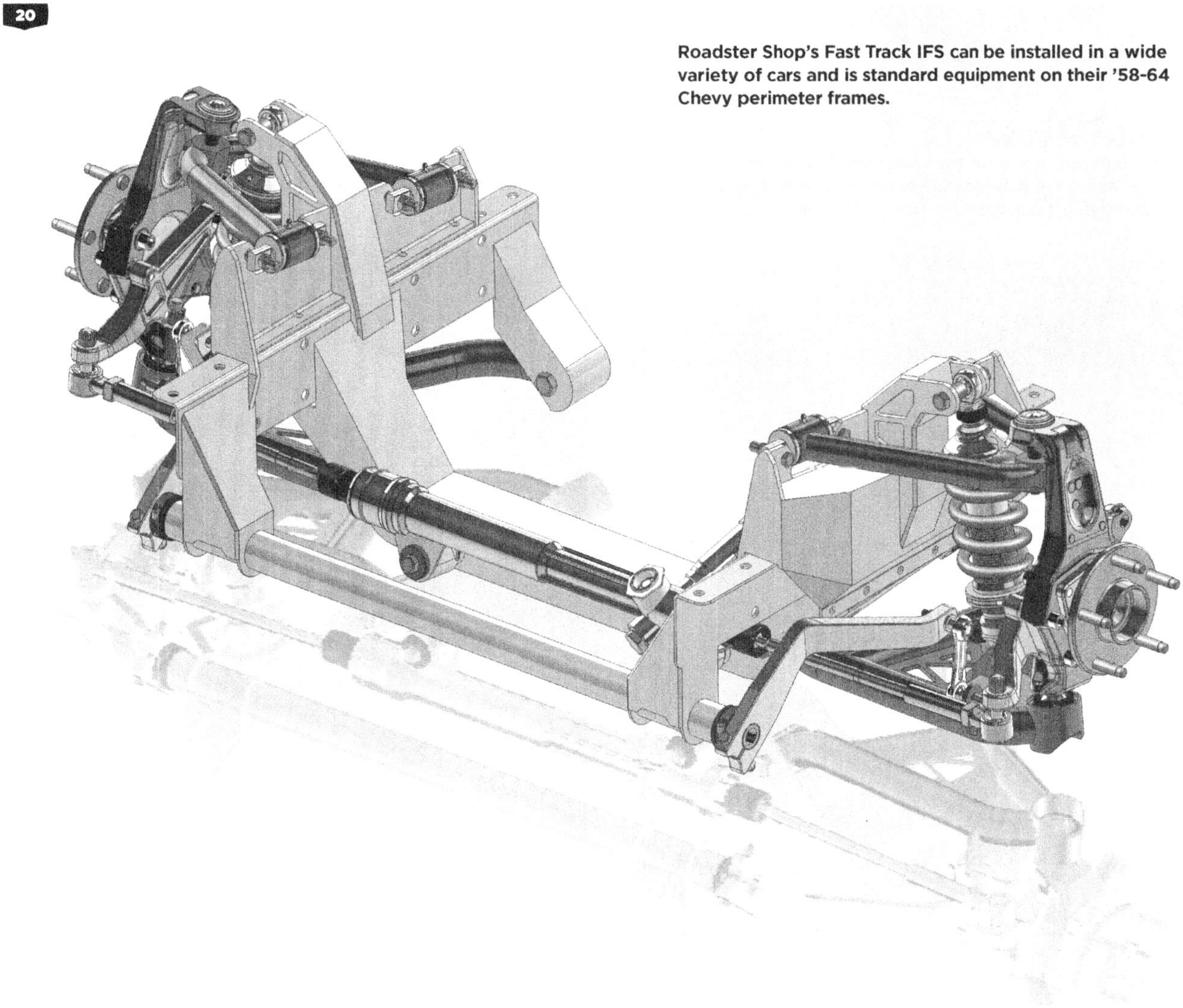

Roadster Shop's Fast Track IFS can be installed in a wide variety of cars and is standard equipment on their '58-64 Chevy perimeter frames.

The Fast Track upper control arms are made from 1-1/4-inch tubing. The upper and lower pivot points are equipped with machined and serviceable Delrin bushing for less deflection and a long service life.

Caster and camber adjustments are established with shims between the upper pivot pins and the mounting brackets. Note the quality of the welds, which are typical of the entire chassis.

For lower unspring weight while maintaining strength, C6 Corvette aluminum uprights are fitted. Hubs and wheel bearings are also Corvette.

Custom steering arms are made from aluminum—Roadster Shop's attention to detail is evident in the machine work.

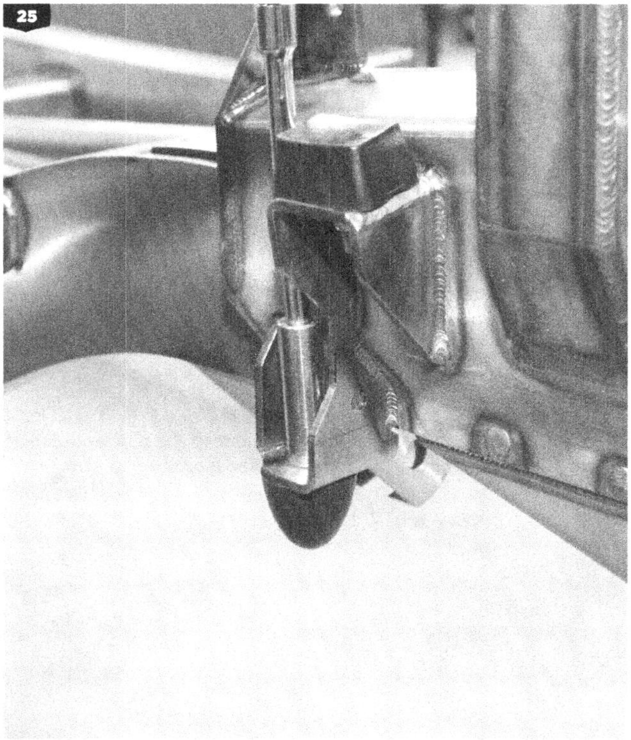

No details have been overlooked—rubber snubbers to cushion maximum compression and extension travel are included.

Custom splined arms attach to the antiroll bar and are secured by pinch bolts.

A splined 1.25-inch antiroll bar ensures our Chevy will stay flat in the corners. It's housed in a tube that is built in the frame and rides on Delrin bushings.

We opted for the double-adjustable RS by Penske shocks. Note the adjustment knobs at the top and bottom of the shock.

29

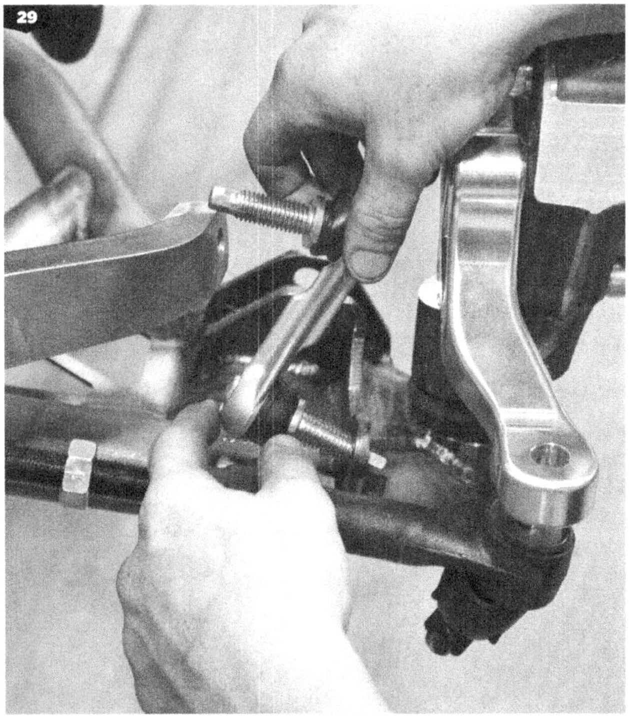

Corvette links are used to attach the antiroll bar arms to the bottom A-frames.

30

Penske is one of the biggest names in racing as well as the production of the high-performance shocks—which are used on all RS Fast Track suspension systems. Note that spherical bearings are used in the mounts rather than urethane bushings.

31

Roadster Shop adds a unique reinforcement to the rack-and-pinion mounts for added strength and to stabilize the steering gear.

32

We used a Billet power steering rack from Flaming River. Made from all-new components, including a new valve and piston, these racks provide over 6 inches of travel.

33

Internally threaded outer tie-rod ends provide toe adjustments. Once set, the tie-rod ends are secured with lock nuts.

The completed Roadster Shop Fast Track IFS is impressive in looks and performance.

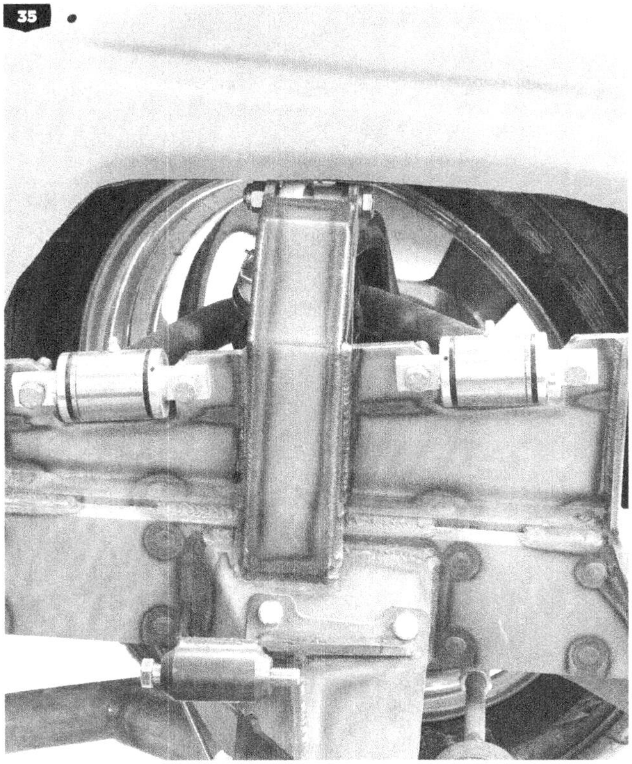

To accommodate the new front suspension, Hot Rods by Dean modified the inner fender panels.

The design of the lower control arm bracket is not only simple and strong, it leaves plenty of room to route exhaust pipes. The engine mounts are for the LS that's to come.

Connecting the steering wheel to the rack-and-pinion steering will be a column from Flaming River.

CHAPTER FOUR

Bringing Things To A Stop

Four-Wheel Discs

By Ron Ceridono | Photography by Brian Brennan

If you think the brakes on your street rod are abused it's nothing compared to the torture tests Wilwood puts their products through.

When our '59 Chevrolet hit the showrooms it was advanced in a number of areas. The styling certainly pushed the envelope, from the "jet plane"–inspired intakes in the hood to the broad fins and "cat's eye" taillights in the rear. Underhood performance options included a fuel-injected small-block or a 348, and either engine could be found backed by a four-speed transmission. But while there were a number of factors that make '59s really cool to this day, driving one makes them less so. Compared to contemporary cars the Chevy has all the handling characteristics one would expect from a car of this vintage—the ride is soft and squishy, which results in so much body roll in sharp corners it seems as though the rocker panels are dragging on the ground. And in today's traffic the brake's shortcomings are evident; bringing 3,500 pounds of Impala to a halt from freeway speeds with the original drum brakes is an eye-opening experience.

To ready our AMSOIL/*STREET RODDER* Chevy for the 2014 Road Tour, and to perform at a level we deemed necessary, Hot Rods By Dean (HRBD) swapped the stock chassis for a Roadster Shop perimeter frame with front and rear independent suspension. With the ride and handling issues addressed, the next consideration was updating the brakes. As we have in the past, we turned to Wilwood for their expertise.

Wilwood began doing business in 1977 and quickly became one of the predominant aftermarket brake companies supplying the race and street/performance markets. Wilwood brakes can be found in NASCAR, drag racing, dirt track, and just about every other form of motorized competition. In addition to motorsports Wilwood also manufactures brake systems and components for agricultural equipment, industrial applications, utility vehicles, snowmobiles, quads, motorcycles, and military vehicles. Of course they also have a vast array of

Dean Livermore of Hot Rods By Dean (HRBD) does a test-fit of the a Wilwood rear rotor and caliper.

With large-diameter brakes wheel fitment can be an issue, these beauties are 19-inchers from Circle Racing Wheels; in the rear 20-inch wheels will be used. Wilwood has fitment diagrams for most brake kits on their website.

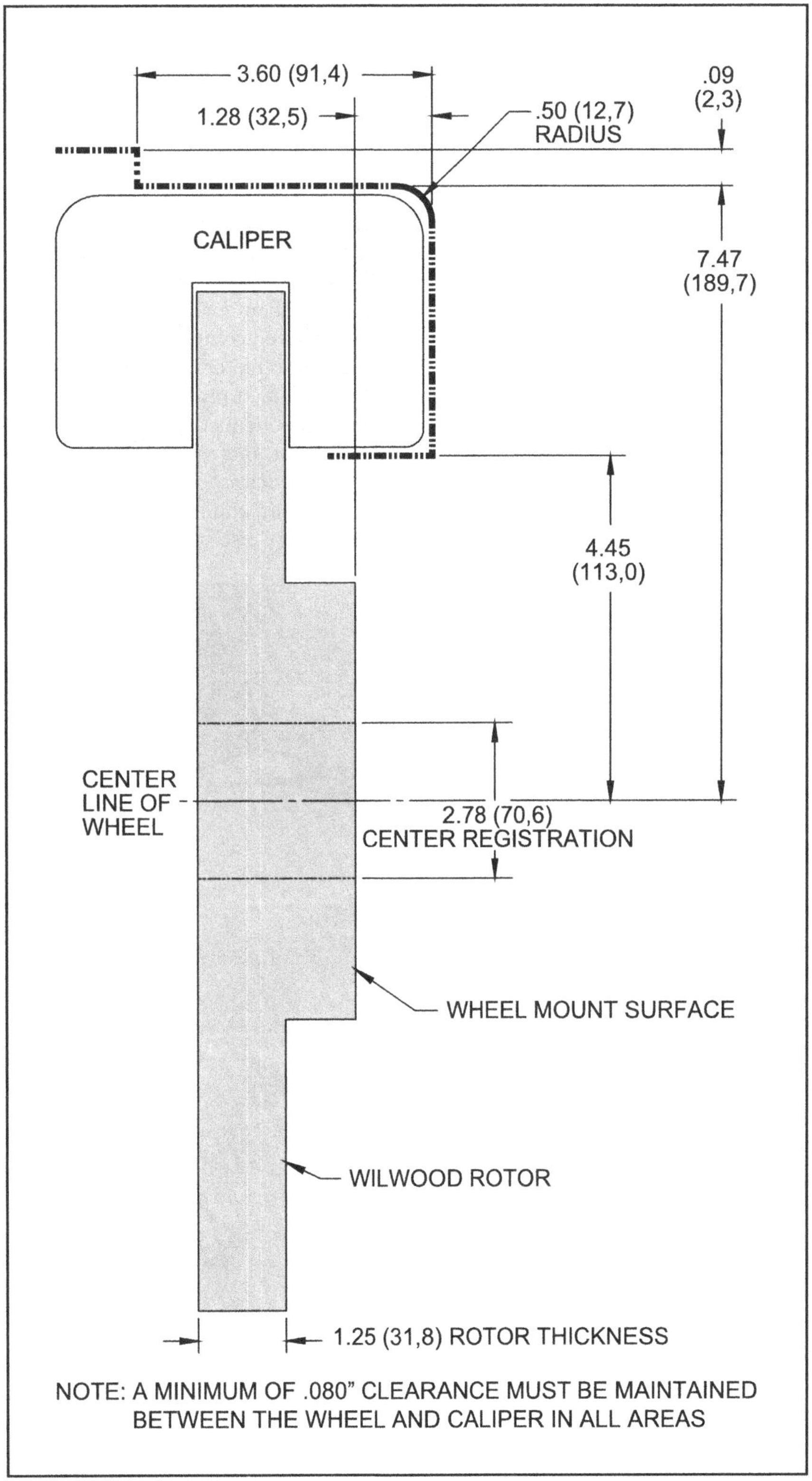

To aid in wheel selection, this diagram from Wilwood gives the dimensions of the 140-8921 brake kit.

The front rotors are certainly impressive looking, note the chamfered holes, the directional slots, and the ends of the screws that hold the rotor to the hat.

street rod components and kits.

The Roadster Shop chassis under our '59 uses C6 Corvette spindles up front, so we chose Wilwood's brake kit for that application (PN 140-8921). Included in the kit are massive Billet Superlite 6R six-piston calipers (PN 120-11778/79). The fully CNC machined billet bodies have six stainless steel pistons (three per side) with

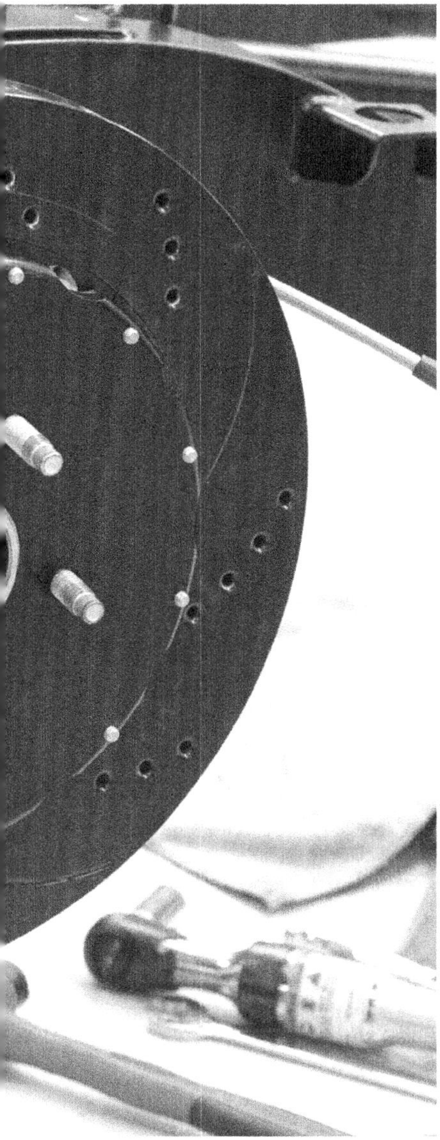

Securing the front hubs to the hats are 12-point screws that also act as the wheel studs. The hubs are drilled and tapped for 5-on-4.5 and 5-on-4.75 bolt circles.

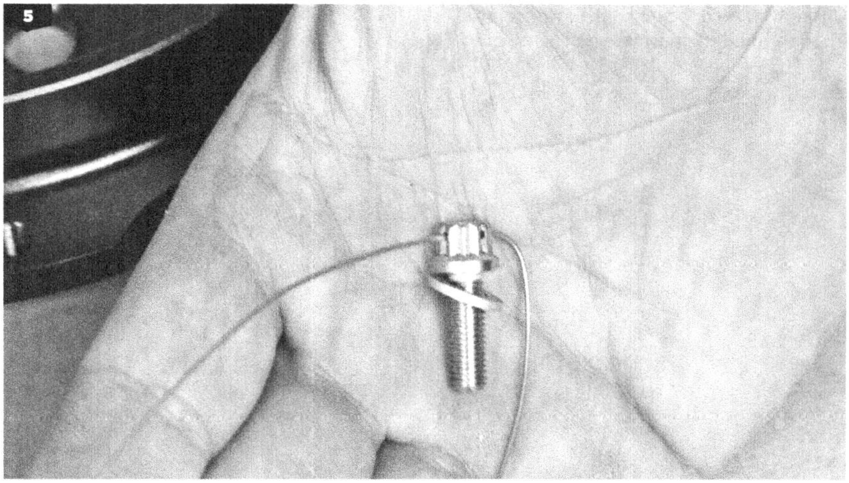

These are the 12-point screws that secure rotors to the hats. They are drilled to accept safety wire.

what is called a differential bore configuration. Adapted from Wilwood's racing brakes, it means that the piston sizes are staggered, in pairs, along the length of the caliper body. According to Wilwood, "Clamping pressure is proportionately distributed to provide balanced pad loading with even wear properties at all loads and temperatures. Pad life is extended and performance is measurably improved." These new versions of the venerable Superlite caliper series

BRINGING 3,500 POUNDS OF IMPALA TO A HALT FROM FREEWAY SPEEDS WITH THE ORIGINAL DRUM BRAKES IS AN EYE-OPENING EXPERIENCE.

are available with radial mounting provisions. This style of mount allows the caliper to be adjusted on two planes—closer to or further away from the centerline of the spindle and in or out in relation to the center of the rotors' braking surfaces.

Rotors in the kit are 13.06x1.25-inch thick, drilled, slotted and carry PN 160-7798/99. As Wilwood describes them, "These rotors feature a specially engineered directional cross drill and face slot pattern that

6

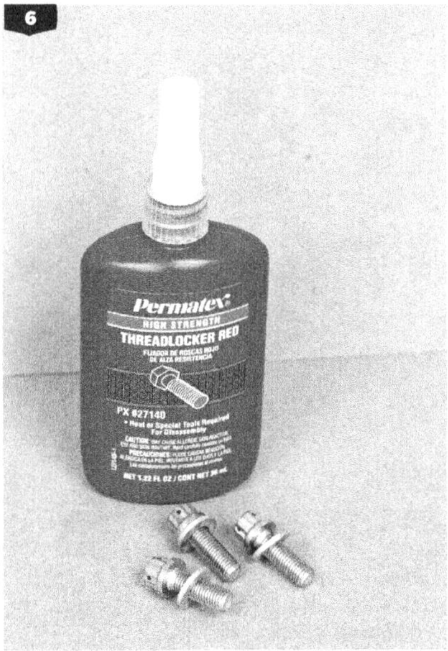

During assembly HRBD applied high-strength thread locker to the rotor screws then torqued them to Wilwood's specs.

7

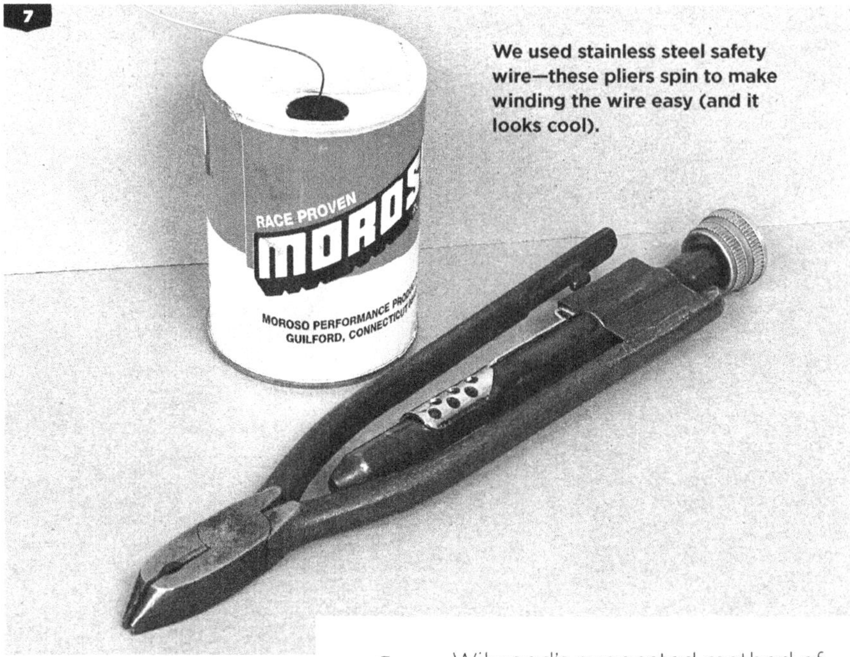

We used stainless steel safety wire—these pliers spin to make winding the wire easy (and it looks cool).

Wilwood's suggested method of safety wiring the rotor screws.

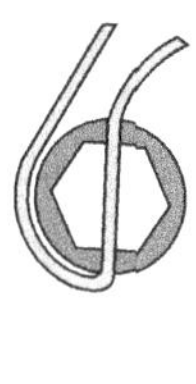

Begin by sliding the 0.032" diameter wire through two of the holes (left) that are 180° apart. Twist the wire as shown (below) using safety wire pliers. Now slide one wire through to of the holes (180° apart) and wrap the other wire around the bolt twist the wires together to form a pigtail. See DS-386 for complete details.

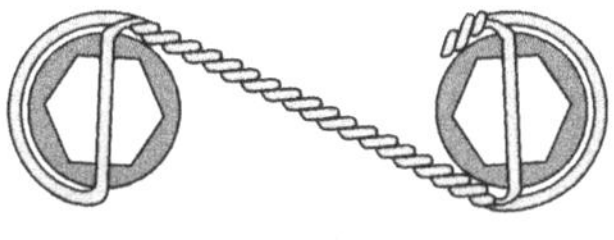

8

Wilwood shows fasteners wired in pairs; safety wire is put through one fastener, twisted, then put through the next one and secured.

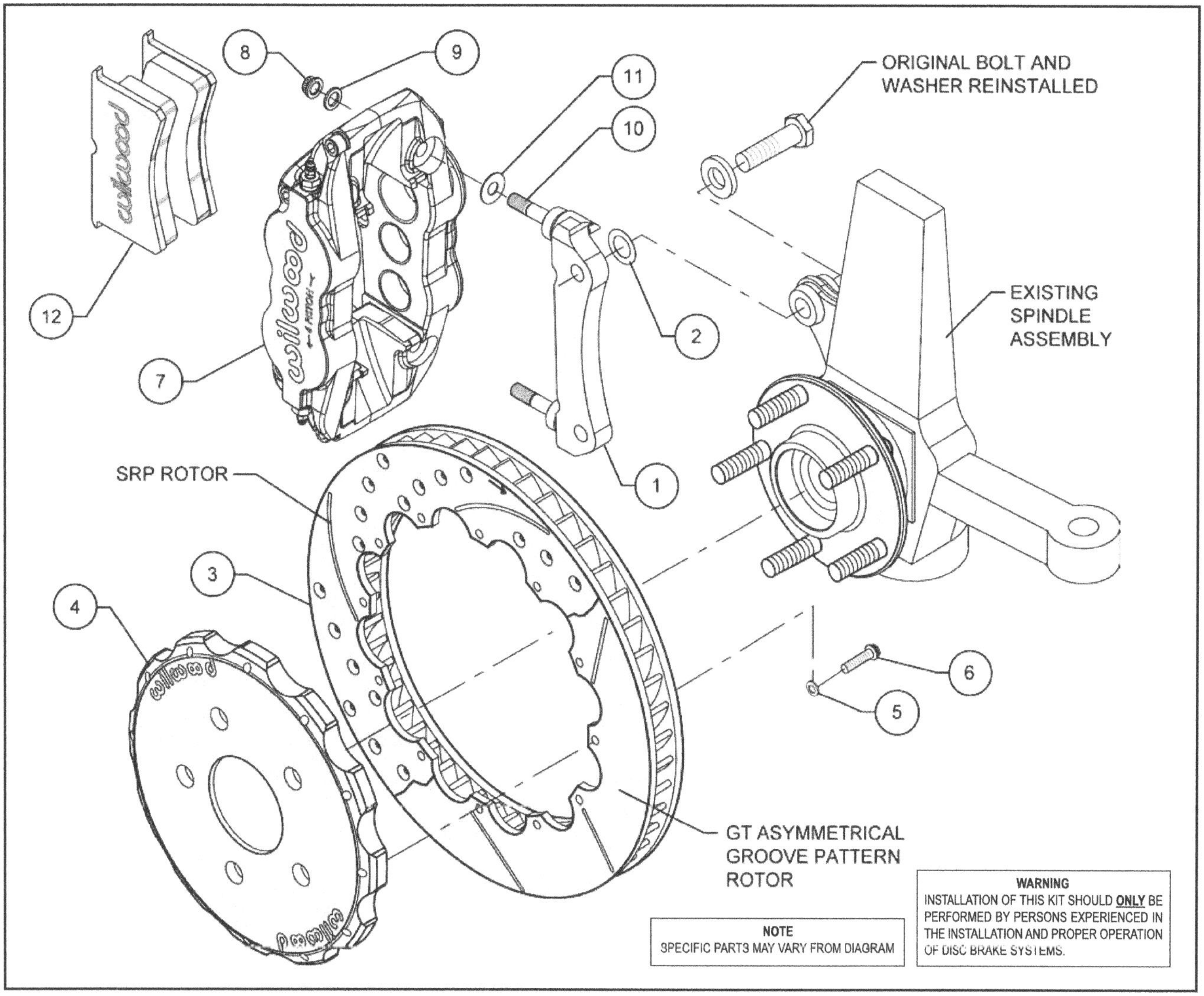

In this exploded view the staggered bores of the front calipers can be seen.

9

The hat for the rear rotor incorporates a small drum for the parking brake. Note the rear rotors are slotted and drilled to match the front.

10

Like the front rotors, the rears' fasteners had thread locker applied, were torqued to specs, and then safety wired.

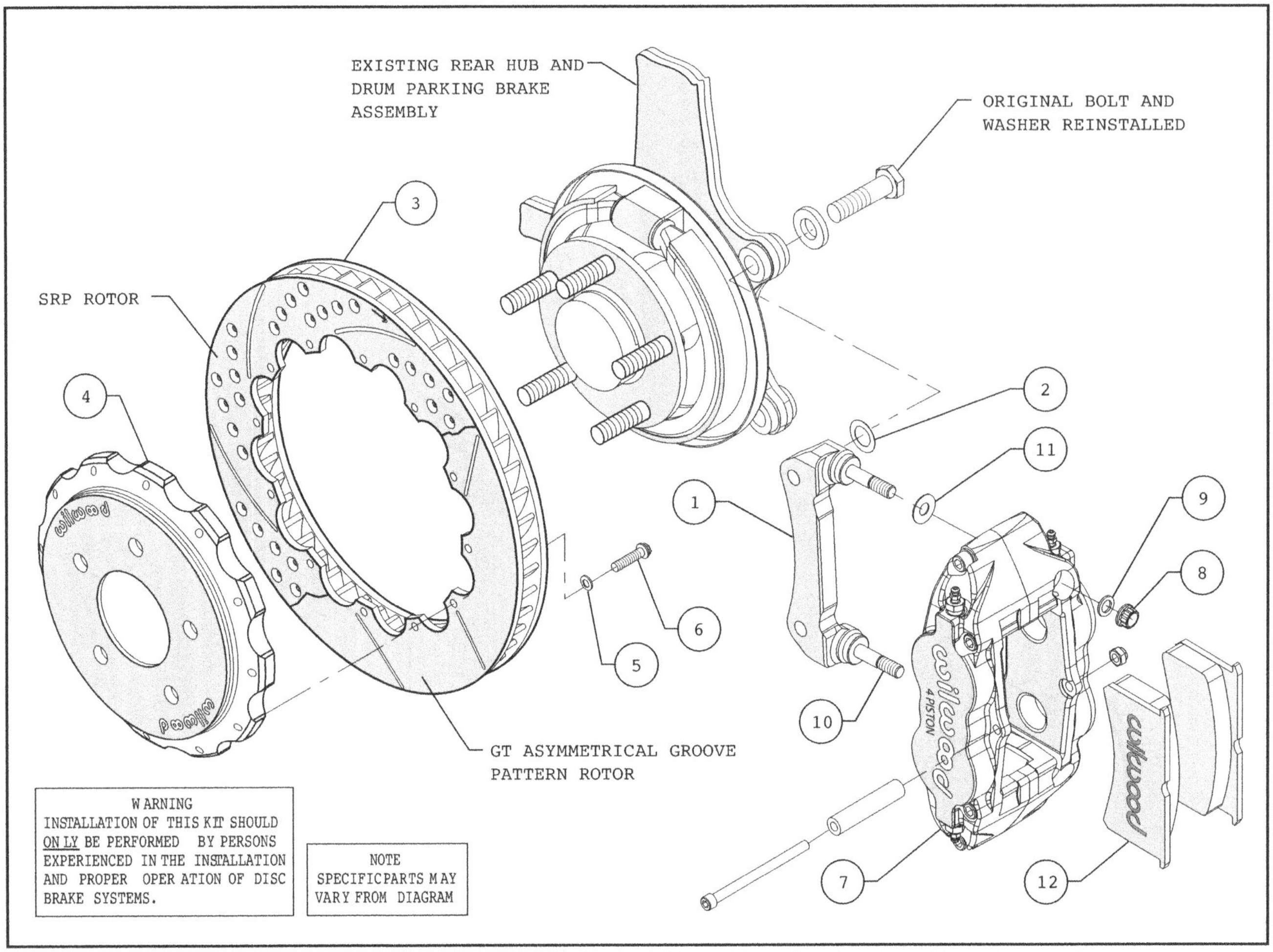

This is an exploded view of the rear brakes. Note the parking brake assembly and the caliper has four pistons, two per side.

11

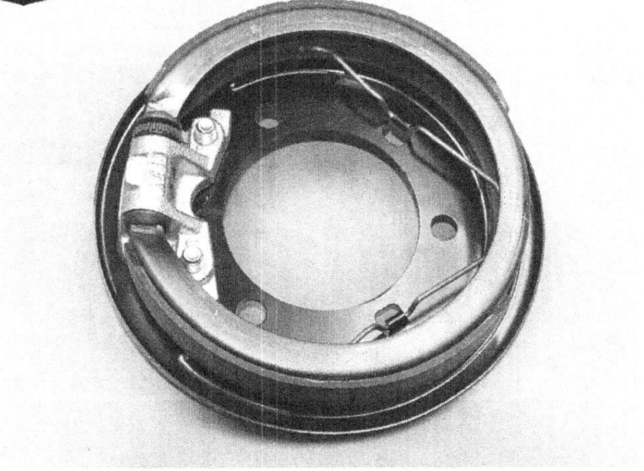

This is the OEM Corvette parking brake assembly from the drum side. Note the adjuster and that both friction surfaces are attached to one steel shoe.

12

From the back, the bracket for the parking brake cable and the actuating lever can be seen.

This is Wilwood's radial caliper mounting bracket. It bolts to the spindle assembly; the two studs mount the caliper.

By placing shims between the mount and the spindle, the caliper can be centered precisely over the rotor.

improve brake response and pad performance throughout the entire range of light to heavy braking. The venting and cleaning action of the holes and slots will reduce pad glaze and disperse gases and heat generated during the pad-to-rotor interface. The holes and slots also inhibit irregular pad compound buildup on the rotor faces resulting in smoother engagement and improved pedal response in all conditions. The material removed also contributes to lower rotating and unsprung suspension weight."

The Roadster Shop's IRS also uses some Corvette components in the rear, namely the hubs and carriers. All the brake components for that application were found in Wilwood's rear brake kit for C5/C6 Corvettes (PN 140-8032). It includes Superlite four-piston calipers that are also made from billet aluminum. These four-piston calipers are said to have the highest clamping efficiency and the lowest deflection of any caliper in its class and are equipped with radial mounts. For this particular application Wilwood's unique rotor hats with integral drums (PN 170-8169) were used to accommodate the stock Corvette internal expanding parking brakes.

A special feature of both the front and rear brake kits are the two-piece hats and rotors that are held together with a series of 12-point screws. These fasteners are drilled for safety wire and it is critical they are secured in the fashion shown here along with a quality thread-locking compound.

Dean Livermore and the crew at HRBD in Phoenix have made amazing progress on the AMSOIL/*STREET RODDER* '59 Chevrolet. With the new Roaster Shop chassis there's no doubt the Impala's ride and handling will be superb, and with Wilwood discs at all four corners, bringing the Chevy to a stop will require considerably less real estate than the original drums required.

Safety Wire Tips From Wilwood

- Using standard (0.032-inch-diameter) stainless steel safety wire, begin by sliding the wire through two of the holes that are 180 degrees apart. The twists should be six per inch, minimum, with a maximum of 12 twists per inch. Also, only safety wire bolts in pairs. Do not string along.

- Note: Safety wire (also known as lock wire) is used to prevent a fastening device from coming loose. It is not a substitute for a secure fastener.

- Safety wire must always tend to tighten the bolt. For a bolt or screw that is tightened by turning it in a clockwise direction, the installed safety wire must tend to pull in a clockwise direction. Select the bolt head holes that would best produce this tightening effect.

- Safety wire must be tight when installed.

- Safety wire must be the proper size.

- Safety wire must never be over stressed. Care must be taken when twisting safety wire together or pulling it tight, not to tighten it so much that it will break under even a light load. Making sharp bends or kinks in the wire must be avoided because they produce stress points in the wire and could cause it to break.

- Always use safety wire pliers to twist the safety wire.

- Safety wire ends must be secured. When installation is complete, the remaining ends are twisted together to form a pigtail. It is standard procedure to wrap the pigtail around the part being safety wired and tucked down out of the way.

  Begin by sliding the wire through two of the holes that are 180 degrees apart. Twist the wire as shown using safety wire pliers. Now slide one wire through two of the holes (180 degrees apart) and wrap the other wire around the bolt. Twist the wires together to form a pigtail.

- Torque values must never be changed to obtain an ideal safety wire position!

- Never reuse safety wire. If you make a mistake, throw that piece away and try again. Safety wire is cheap.

15

Here a six-piston front caliper is being installed; by adding or subtracting shims from the studs the pads can be located properly in relation to the rotor's friction surface.

16

Brake pad changes are made by simply removing two retainer pins, sliding the old pads out and the new pads in, and then replacing the retention hardware.

17

Our Roadster Shop chassis came plumbed for brake lines, but for those who fabricate their own consider investing in an Eastwood flaring tool. It makes forming leak-free flares in steel and stainless steel lines quick and easy.

18

Eastwood's tool can be used to make 37- and 45-degree flares.

19

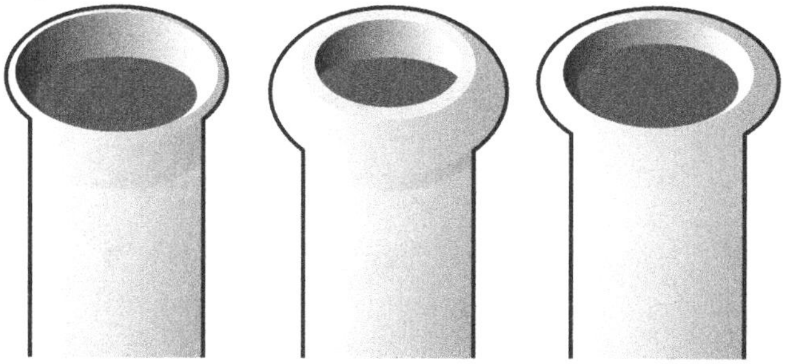

The common types of flares can be made with the Eastwood tool—37 degree for AN fittings, bubble flares found on many imports, and 45-degree double flares.

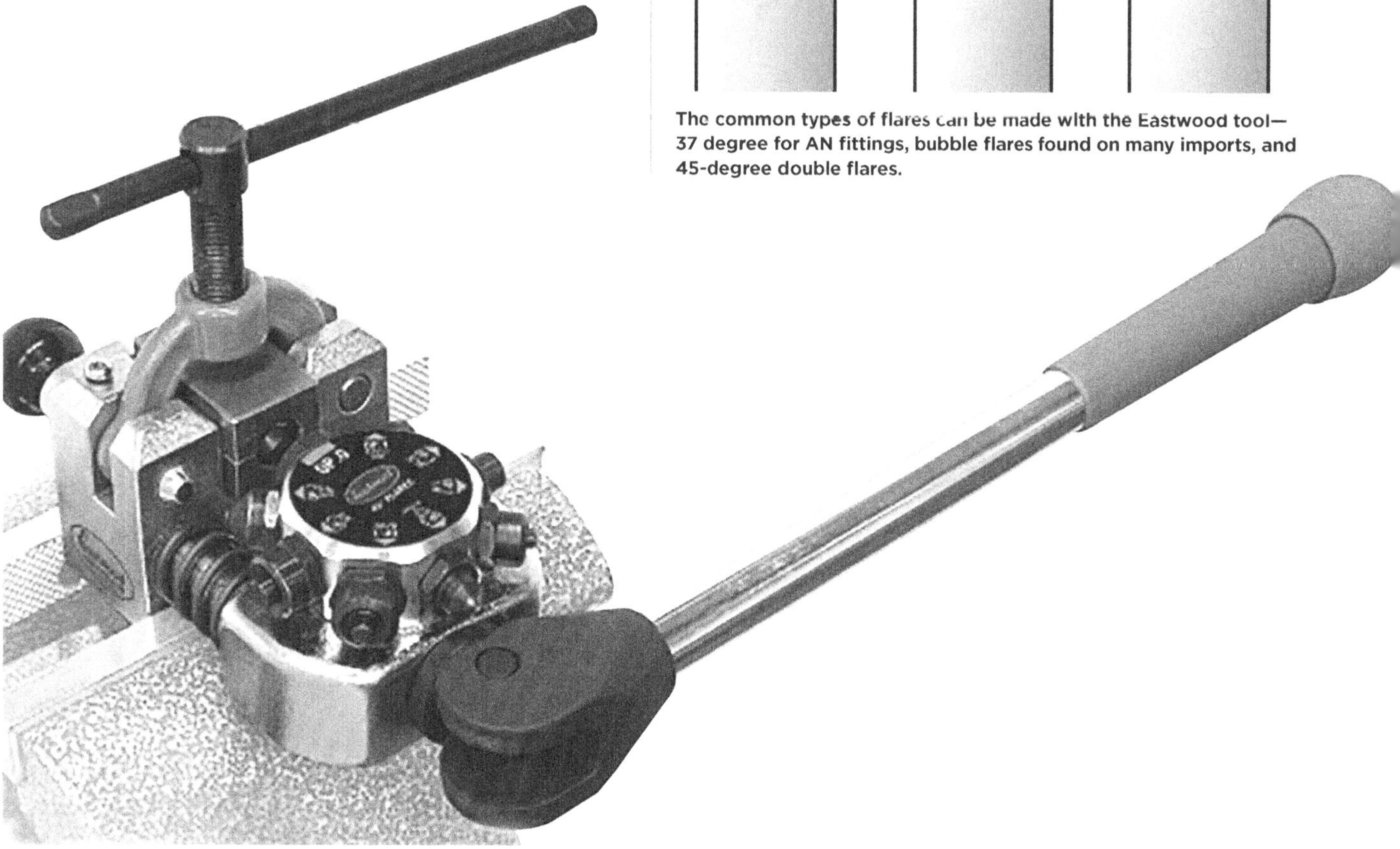

CHAPTER FIVE

When is Enough, Enough?

Shaver Racing Engines Builds a Streetable 622hp 427 LS

By Jeff Smith | Photography by Brian Brennan and Randy Lorentzen

It's big. When the decision was made to build the 2014 AMSOIL/*STREET RODDER* Road Tour '59 Chevy Impala—everything had to be upsized. The car is monstrous, the trunk can easily accommodate a small college marching band (minus the tubas), and the wheels and tires are proud of their illusions of grandeur. So it goes without saying that the engine would have to be equally as momentous. That inferred a big-block Chevy or perhaps a W engine.

But this is 2014, not 1965. So something new means LS power with EFI; the call went out to Ron Shaver (the patriarch of Shaver Racing Engines). He said he could build a 427 that would fulfill all the prerequisites of big power, be very streetable, and still be enticing to the eye. We're here to tell you this big dog hunts on all counts. The old standard for a street performance engine used to be 1.1 horsepower per cubic inch (hp/ci). This RHS block–based 427 blows that out of the water with an impressive 1.45 hp/ci with its 622 hp peak. Even better, this pump gas performer does it with respectable idle quality, outstanding throttle response, and a power curve that will demand you flatten that skinny little pedal on the right as often as possible. Diplomatic immunity might be a requirement to drive this car.

All this magical power didn't happen accidentally. The first rule was to not cheap out. Shaver could have used a production block, but the commitment to quality demanded the near-legendary RHS aluminum block as its foundation. We combined the RHS block's cornerstone with a durable rotating assembly from SCAT, Mahle pistons, an outstanding set of CNC-ported heads from Livernois, an over-achieving valvetrain from COMP Cams, and an absolutely stunning Inglese 8-Stack induction system to come up with the ultimate street engine. Please excuse us if we gush a bit, but a mild big-block Chevy 502, for example, can make an impressive 570 lb-ft of torque at 4,200. The Road Tour 427 thumps out over 600 lb-ft at 4,800 rpm and does it while trimming 200 pounds. And we're doing it in style. Let's see how our Road Tour 427 all came together.

BUILDING BLOCKS

It wasn't long after the LS engine made its debut in 1998 that Corvette escalated the horsepower game with a 4.125-inch bore 427 dubbed the LS7. Since then, the 427 has become the *de rigueur* displacement for powerful LS street engines. While GM sells a 427 production block, for the Road Tour '59 we opted for the top-of-the-line RHS aftermarket block. The block comes in several configurations but Shaver

(Many pretty engines are just that–attractive but a little short on real power. The AMSOIL/*STREET RODDER* Road Tour LS 427 is a rare combination of statuesque good looks combined with eyeball-flattening power. At 620-plus horsepower, Dean Livermore of Hot Rods by Dean is making final adjustments, after dropping it into the Road Tour '59 Chevy he thinks it will get the job done.

spec'd the standard deck height 4.125 finished bore assembly that would also easily accept the 4.00-inch stroke crank necessary to complete the 427 combination. The RHS block includes longer cylinder sleeves to accommodate the longer stroke, which avoids the difficulties stemming from shorter cylinder lengths in the LS2 and LS3 OE blocks.

RHS blocks also feature a 0.388-inch raised cam bore centerline to ensure it can clear even the longest stroke crank. That requires a longer chain that COMP provides. The cross-bolted mains and deep skirt block also offer increases windage passages to minimize windage difficulties that can occur with longer stroke LS engines.

Combined with a priority main lubrication system with reduced restrictions, this block can easily accommodate much bigger power numbers, which means it promises to make power for a very long time.

The rotating assembly is also up to the challenge with a SCAT forged steel 4.00-inch stroke crank and I-beam forged 6.125-inch connecting rods that Shaver fitted to a set of Mahle flat-top 4.125-inch bore pistons with matching Mahle rings. The cam is a COMP hydraulic roller with mildly aggressive

1

We needed a strong foundation for the Road Tour 427 and it doesn't get much better than an RHS standard deck height alloy block. The RHS block comes with priority main oiling, which means the mains and rods get oil first. The block also has a raised cam centerline to clear big stroke cranks.

2

A big advantage with the RHS block is spun cast-iron liners that ensure excellent ring seal. We didn't need them, but you can see that the block can accept six-bolt cylinder heads. A standard deck RHS block can accommodate a stroke up to 4.250. With this 4.125-inch bore, that's 454 inches.

3

A forged steel crank is the only choice for an engine like this, so we selected a SCAT forged steel crank and matched it with a set of SCAT I-beam rods.

4

Mahle makes an excellent 4032 alloy flat-top piston for the 427 that includes a nice coating on the skirts. The 4032 alloy does not expand as much as 2618 alloys for a tighter piston-to-wall clearance. This makes for a quiet engine during warm-up.

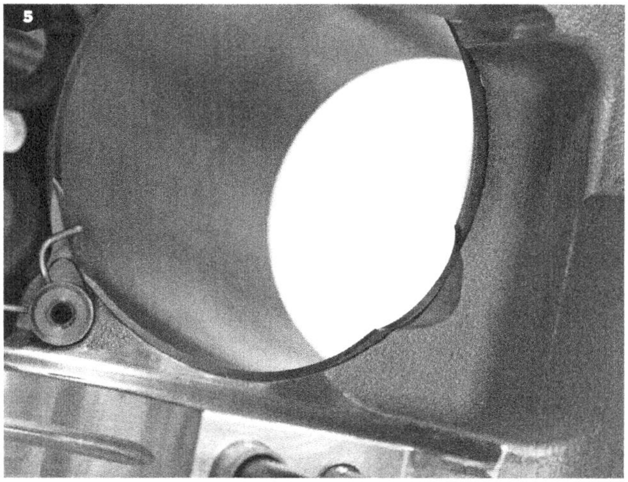

5

An initial mockup by Shaver engine builder Lee Clapp revealed that the connecting rods required minor clearance on the bottom of the cylinder sleeves. The RHS piston oilers also had to be modified to work with the longer stroke.

6

Pre-assembly includes setting end ring gaps on the 0.043 top and 1.5mm second rings. The oil rings are 3 mm. Conventional wisdom now dictates a slightly wider second ring gap to prevent pressure building underneath the top ring, which can unload it causing leakage.

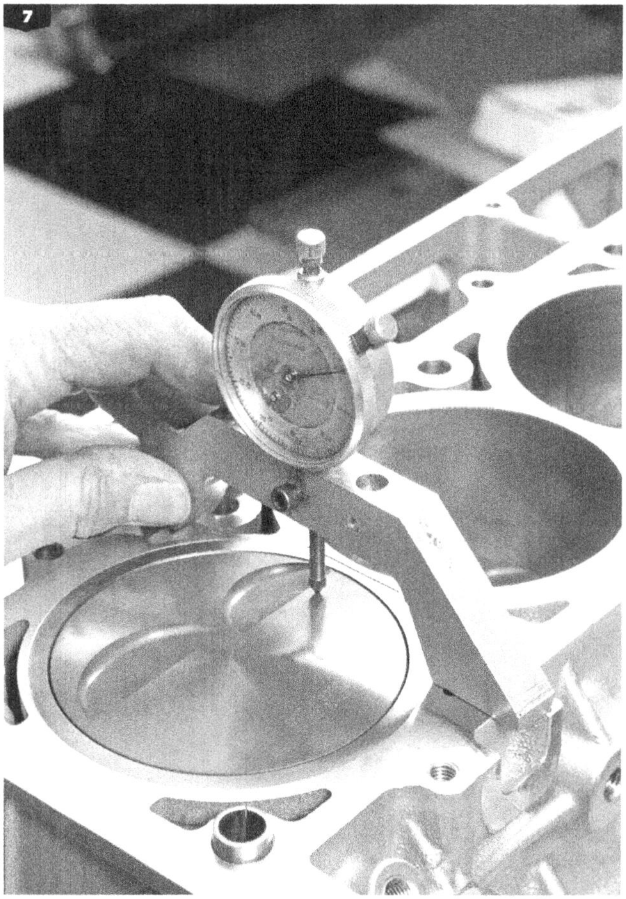

7

Checking the piston deck height is also crucial to compute the static compression ratio. The Mahle flat-top pistons protrude 0.006 inch out of the top of the block, which is fairly typical for most LS engines when using thick MLS gaskets.

specs (see chart) but even when combined with the rectangle port heads by Livernois, this engine doesn't sacrifice torque to make horsepower.

The heads are worth investigating more closely. Livernois starts with a pair of 821 LS3 rectangle port production castings. These are slightly different than the L92 rectangle port truck heads. The LS3 has a slightly improved intake port floor with a roughly 260cc intake port volume. Livernois applies a CNC porting effort to both the intake and exhaust ports, opening the intake up slightly, which produces a peak lift flow of 360 cfm, which is roughly a 10 percent flow improvement over stock. What is often overlooked with cylinder heads is how well the exhaust port performs. The exhaust cranks out a peak number of 245 cfm (275 if a 1-3/4-inch primary header pipe is used), which generates a very efficient 76 percent intake to exhaust ratio. This enhanced exhaust flow often contributes to excellent high-rpm power numbers. Combined with the Livernois heads is a set of Manley 2.160 intake and 1.590-inch exhaust valves that are basically the stock sizes. COMP then supplied a complete set of LS3 shaft-mount roller rocker package that uses an 8620 hardened steel shaft to create

CAM SPECS				
XER281HR-12	Adv. Duration	Dur. at 0.050	Lift (inches)	Lobe Sep.
Intake	281	232	0.595	112
Exhaust	283	234	0.598	

Once the block was finish honed to fit the Mahle pistons and subjected to a good cleaning, Clapp dropped the crank in for the first of several test-fits. Bearing clearances, crank endplay, rod side clearances, and a dozen other critical dimensions all have to be checked before final assembly. This time-consuming effort is why it there are no shortcuts to a professional build.

After checking wrist pin clearance (0.0015 inch is a good number), Clapp pushed the round C-clips in place. This is best done only after the pre-assembly work is complete in case minor modifications are required.

COMP supplied the entire valvetrain, including the XER281HR hydraulic roller cam and the retro-fit short-travel roller lifters. These lifters require precise pushrod length and accurate preloading but perform virtually like a solid roller lifter even at 7,500 rpm.

Installing the cam with a smidgen of lube is sufficient since the engine will be pressure lubed before firing it for the first time.

The RHS block raises the cam centerline, which requires a custom timing chain with two extra links to maintain the proper tension. This was followed by the oil pump attached with ARP fasteners.

A balancer plays a critical role in damping natural crankshaft oscillations. The beauty of ATI's three-piece design is that you can match each piece specifically to your engine.

Before bolting on the Livernois heads, Clapp employed a set of Fel-Pro 0.053-inch MLS head gaskets to reduce the static compression ratio to 10.6:1.

a very rigid mount for the 1.7:1 aluminum rockers. The roller trunion bearings and roller tips are both lube by an internal oiling passage built into the rockers. Combined with a set of dual springs, the valvetrain is more than up to 7,000-rpm escapades.

After engine builder Lee Clap assembled the RHS long-block, it was time for the induction system. Instead of settling for a carburetor or run-of-the-mill EFI, we opted to outfit our Road Tour 427 with a much more intriguing Inglese 8-Stack induction from Inglese. The intake centers around eight 58mm EFI throttle bodies that are designed to appear much like '60s-era Weber 48 IDA carburetors. Each throttle body is equipped with a 50 lb/hr FAST injector and connected with a complete fuel rail system that is visually downplayed. Buried underneath the intake is a series of internal passages that combine all eight stacks together to create a mini plenum that's used both to dampen the individual pulsations that can wreak havoc with pure individual runners. This common area also produces a convenient central point for both referencing manifold absolute pressure (MAP) and also for the idle air speed (IAS) motor that the computer uses to control idle speed. That way, when you crank up the Vintage Air A/C compressor, the FAST controller can easily compensate for the compressor load and maintain the desired engine idle speed.

This leads us to the brains behind controlling the Inglese 8-Stack. We could have gone right to the most sophisticated FAST controller, which would have been a logical step, but we decided to try FAST's new self-learning EZ-EFI 2.0. You might already know about the basic EZ-EFI 2.0 system that works with a single throttle body that replaces a typical carburetor. But FAST also makes a relatively affordable multi-point

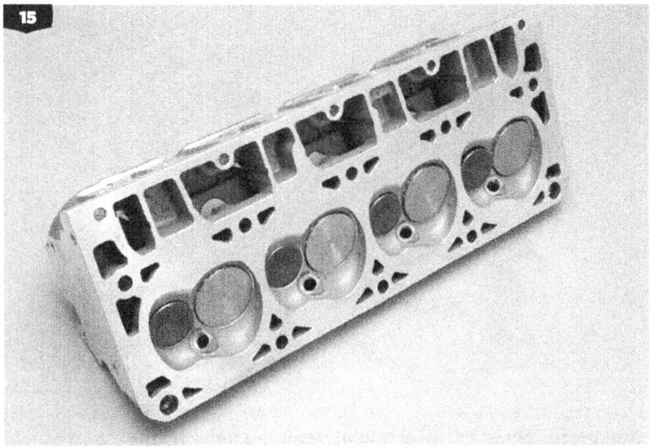

Livernois Motorsports supplied the CNC-ported production LS3 castings for our 427ci. The rectangle port heads are big as-cast, but you'd never know it by how much torque this engine makes.

The Stage 3 Livernois heads run 2.160 and 1.590-inch valve sizes to ensure excellent flow. The chamber size is standard at 70 cc.

Livernois says these rectangle port heads will flow 360 cfm on the intake side and with a 1-3/4-inch primary header tube the exhaust side will flow 275 cfm, which is an exhaust-to-intake relationship of 76 percent, which is outstanding given the big intake flow numbers.

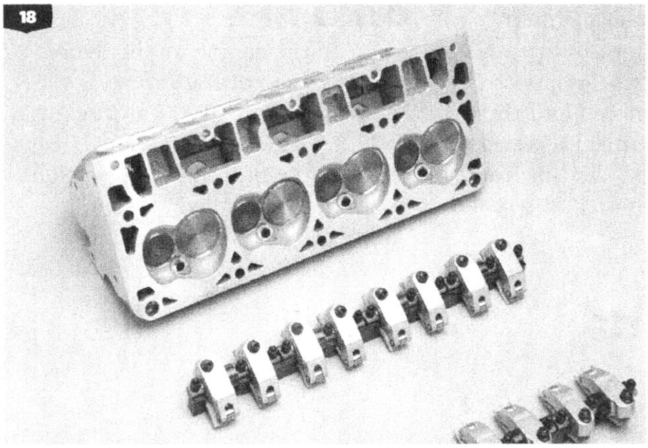

With the large intake valves on the LS3 heads, this moves the valve and requires an offset rocker arm. COMP Cams has a very nice rocker shaft valvetrain for the 427 with the offset intake rocker using the stock 1.7:1 rocker ratio. The shaft offers substantial improvements in valvetrain stability over stud-mounted rockers.

A pro build means degreeing the camshaft to ensure that it is installed at its intended intake centerline. The COMP adjustable timing set made it easy to ensure the 109-degree centerline position.

COMP also supplied the front cover that includes the cam sensor and Clapp bolted it all together with ARP fasteners.

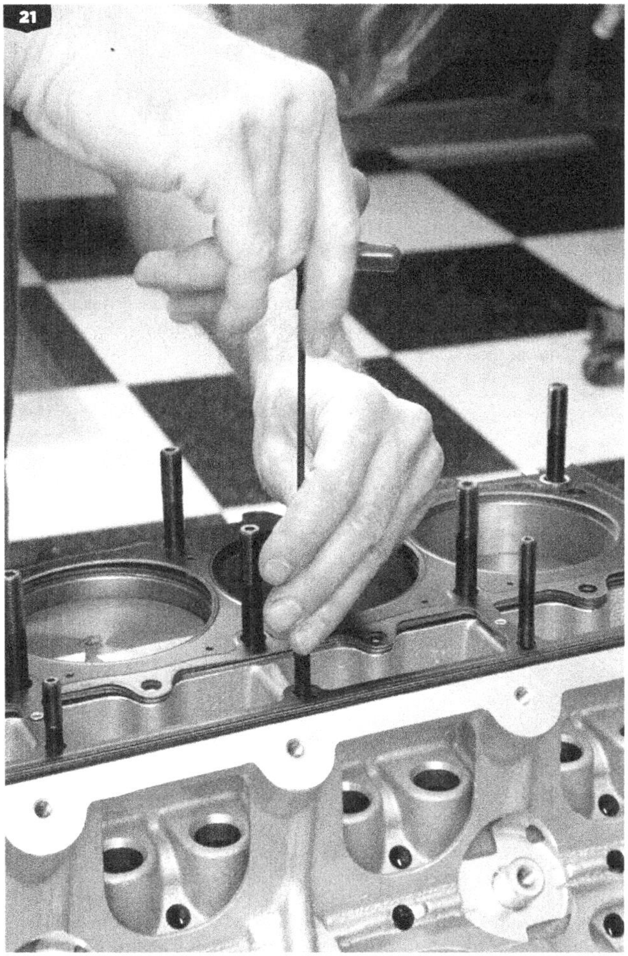

There are lots of reasons to use head studs instead of bolts, including consistent clamp loads. Never torque studs into the block, ideally they should be finger tight with full thread engagement. A portion of the stud thread should show above the deck. This ensures that the top thread in the block is not overly stressed.

system that is designed to work much like the throttle body system but integrates the eight individual injectors and throttle bodies. The best part of the EZ-EFI 2.0 system is its self-learning capability. You set the parameters with idle, cruise, and WOT air/fuel ratios and the system uses feedback from the oxygen sensor in the exhaust to maintain these inputs at all times. It operates as simply as its description. With the EFI system assembled and the Vintage Air Front Runner accessory drive bolted down, it was time to put the 427 on the dyno and test its mettle.

DYNO TIME

With the engine on the dyno, Shaver's dyno operator Ron MacAskill set the WOT air/fuel ratio to 12.8:1 and integrated 30 degrees total timing into the FAST ignition controller and pulled the handle. After several runs up the rpm curve, the Road Tour 427 rewarded

22

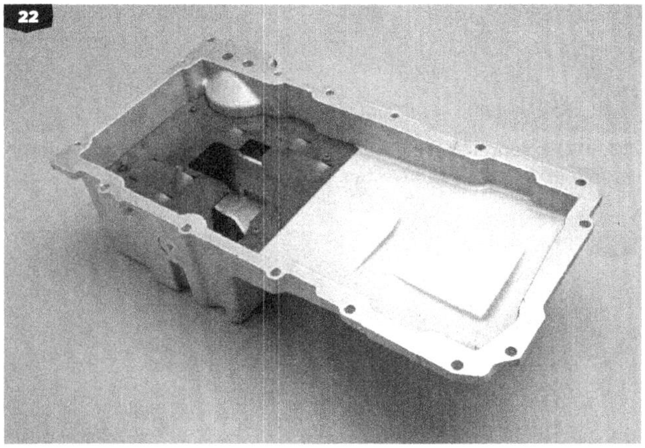

Shaver also decided on a Mast Motorsports oil pan that would work well with the RHS block and also clear the crossmember on the Roadster Shop chassis. The pan is cast aluminum and is intended for use in autocross or road racing applications. Plus, it has a cast-in filter boss, eliminating the need for an external filter.

23

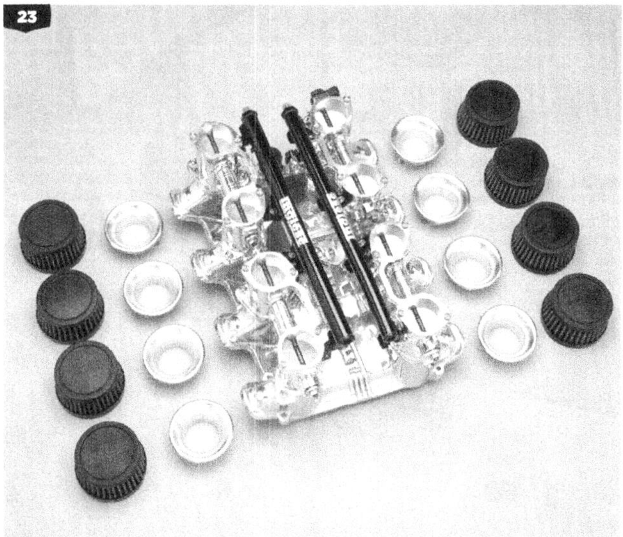

Rather than go with a mundane carburetor or factory intake for the 427, we chose an Inglese 8-Stack induction system. The eight individual 50mm IDA-style throttle bodies each fitted with a 50-lb/hr FAST fuel injector. In addition, the fully assembled manifold is complete with installed fuel rails and an extremely precise throttle linkage system. Intake bells and filters were also part of our assembly.

24

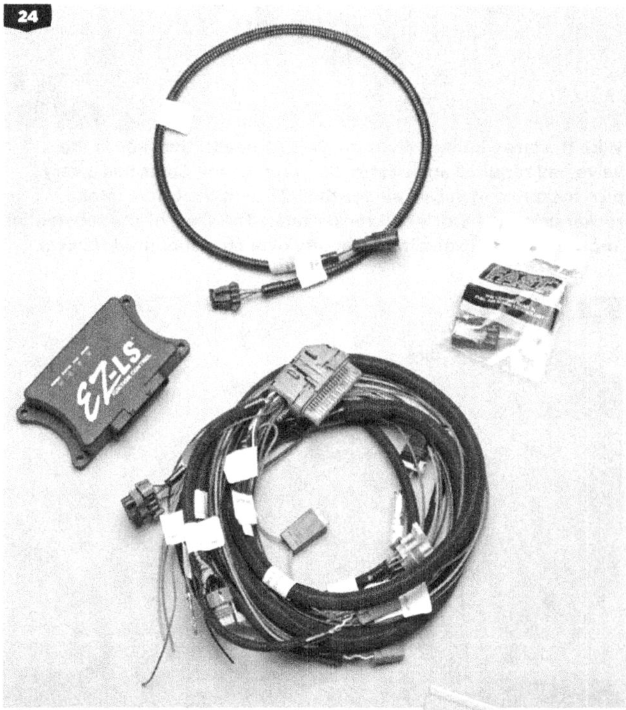

Controlling the Inglese 8-Stack is a FAST EZ-EFI 2.0 self-learning EFI. The beauty of this system is that with just a few keystrokes, the system establishes the necessary parameters and the engine runs and quickly settles into the proper air/fuel ratios.

25

During the assembly of any engine a great choice for engine assembly lube is the by AMSOIL INC. It will stick to components and provide rust and corrosion protection, will dissolve in engine oil, and contains antiwear additives all during the assembly and "storage" cycle before the engine is put into service.

us with a solid 534 lb-ft of torque at 3,500 rpm. Had we pulled the engine down to 3,000, it's likely the big-inch LS will make 500 lb-ft of torque. That's enough to get even a big '59 Impala moving. The power only went up from there with a peak torque of 602 lb-ft at 4,800 rpm and then on to an impressive 622 hp at 6,000 rpm. That's a nice 1,200-rpm power band between peak torque and horsepower and promises to be able to annihilate the tires nearly at will. We can't wait. The next moves might find our intrepid '59 moving under power at last. *

Clapp test-fit the intake on the engine to ensure everything bolted up properly. The Inglese 8-Stack intake uses O-rings instead of gaskets, which makes the installation on an LS engine nearly foolproof. The Inglese manifold also integrates a common chamber for the MAP sensor and idle airspeed controls that make running an individual runner manifold like this much easier. No more balancing throttle blades with a Uni-Syn!

27

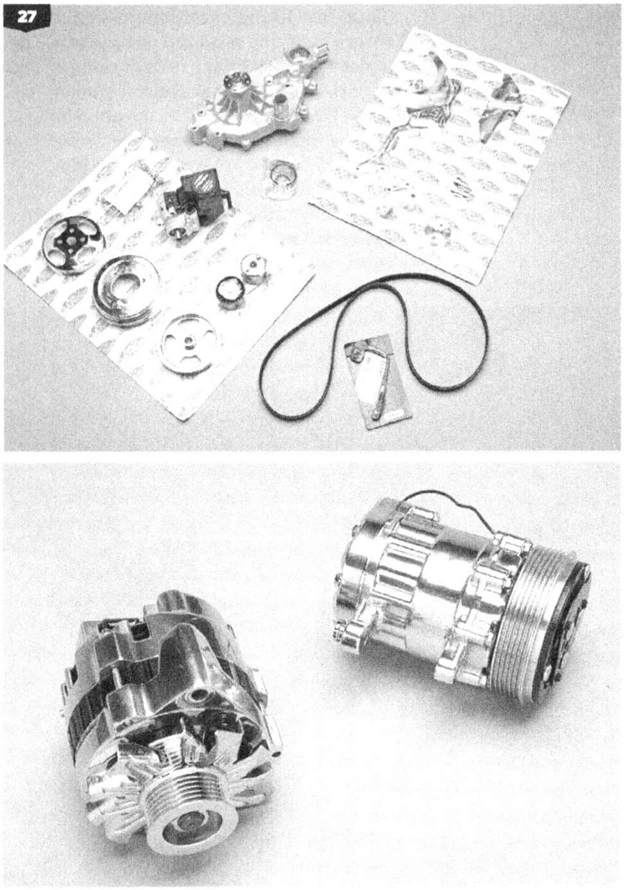

The Vintage Air Front Runner accessory drive assembly comes with everything you need to install, including the water pump, alternator, power steering pump, A/C compressor, and all the brackets and hardware.

28

Flowmaster LS block hugger-style headers (PN 814121) were used, giving plenty of clearance to the Flaming River steering and the Wilwood brake components.

29

Speedway Motors provided the LS finned valve covers while Lokar supplied the engine oil and the trans dipsticks.

30

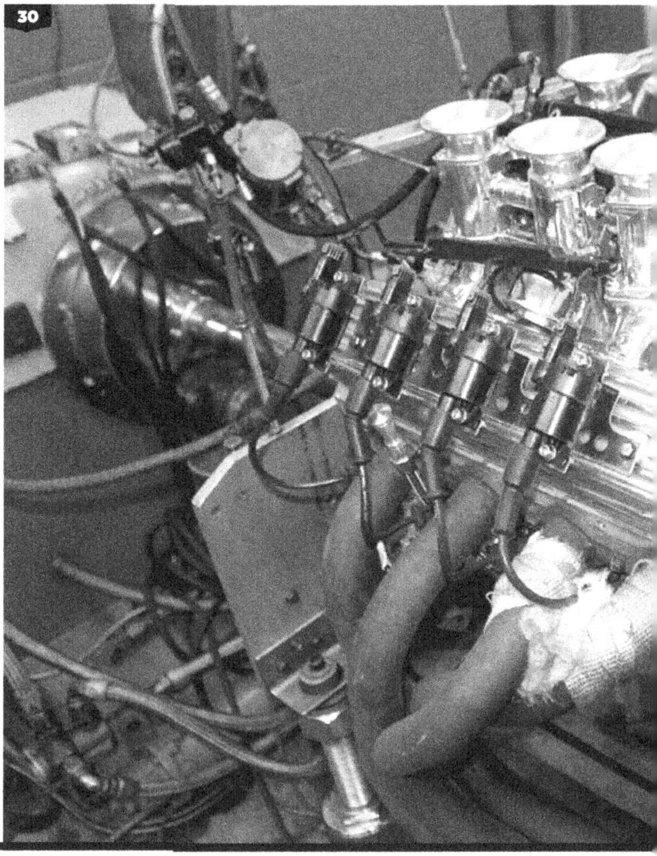

POWER CURVE		
RPM	TQ	HP
3,600	534	366
3,800	543	393
4,000	559	425
4,200	579	463
4,400	596	499
4,600	597	523
4,800	602	551
5,000	601	572
5,200	598	592
5,400	590	607
5,600	579	617
5,800	561	620
6,000	544	622
6,200	520	613
6,400	490	597

It's a good sign when the torque curve on a street engine starts at 500 lb-ft of above. While the test started at 3,500, you can extend the curve and see that even at below 3,000 rpm the engine will make at or nearly 500 lb-ft. That's big-block territory. Then add 620-plus horsepower and it's all gravy.

Installed on the dyno, the EZ-EFI 2.0 system fired the engine immediately and after a short engine break-in period, the engine was ready to impress. It only took a couple of runs for the EZ-EFI 2.0 system to dial-in the air/fuel ratio for peak power, making a best of 622 hp at 6,000 rpm.

PARTS LIST	
DESCRIPTION	PRICE ($)
RHS LS engine block, 9.240 inch	4,699.97
RHS piston oiler kit	597.97
COMP LS front cover kit	240.97
Edelbrock lifter valley cover	119.97
SCAT LS 4.00 stroke crank, 58x	973.97
SCAT 6.125-inch rods	439.97
Mahle PowerPak piston kit	766.67
Clevite LS main bearings	218.97
Clevite LS rod bearings	71.76
Livernois CNC-ported LS3 heads	2,899.99
ATI F-car LS1 balancer 7.5 inch	435.00
Inglese LS3 48IDA system	4,500.85
Inglese stacks, 50mm, 1.50 inch	199.95
Inglese Snap-In filter kit	283.60
FAST 50 lb/hr injectors	403.97
FAST EZ-EFI 2.0, Multipoint	1,249.97
EZ-LS GM Ignition controller	N/A
FAST L-92 coil pack	305.97
Ported LS3 heads	2,899.99
COMP Cams XER cam	394.97
COMP timing set, hex-adjust	299.99
COMP hyd. roller lifters	534.97
COMP LS rocker shaft kit, LS3	1,399.97
COMP dual valvesprings	399.97
ARP head stud kit	554.74
ARP accessory bolt kit	283.16
Fel-Pro MLS head gaskets, left	74.97
Fel-Pro MLS head gaskets, right	75.97
Fel-Pro timing cover set	30.97
Fel-Pro oil plan	32.97
Fel-Pro valve cover	10.97
Fel-Pro header gaskets	8.97
Mast Motorsports oil pan	399.99
Sealed Power oil pump, LS3	99.97
AMSOIL INC. oil filter	13.70
Speedway starter motor	234.99
Speedway LS finned covers	399.99
Vintage Air LS Front Runner	2,694.97
AMSOIL INC. 10W-30 oil	8.97 (5)
Lokar engine oil dipstick	58.97
AMSOIL INC. Engine Assembly Lube	
Champion spark plugs	23.76 (8)
Aeromotive fuel pump	
Flowmaster exhaust manifolds	
Optima Red Top "Large" 4 Post	
LSx/Vortec Thermo Housing Pol	
MSD 8.5mm LS1 Blk. Wire Set (St)	

CHAPTER SIX

Keeping Our Impala Cool

Choosing a Radiator and Fan

By Ron Ceridono | Photography by Brian Brennan

Like its predecessors the AMSOIL/*STREET RODDER* Road Tour '59 Chevrolet is a combination show car and road warrior. Built by Hot Rods by Dean to their typically high standards, our Impala has to look good and still be capable of putting 20,000 or more on the odometer in a single summer. Not only does the car cover a considerable amount of ground traveling to far-flung destinations, the tight schedule leaves zero time for breakdowns.

One of the problems that often plagues modified cars is overheating, so over the years we've turned to Don Armstrong at U.S. Radiator for his guidance in keeping our Road Tour car's operating temperatures in check. Armstrong knows the radiator business—he's been at U.S. Radiator almost from the company's beginning. He began as a delivery driver, worked in every phase of radiator construction, and today owns the company. Under Armstrong's leadership the company produces radiators for over 400 different applications, and if they don't have what you need they can probably make it. To top it off, Armstrong is an avid hot rodder, so when he talks about keeping one cool, we listen. Here's what he had to say on this very important subject:

"While all radiator cores might look the same, they perform vastly different based on tube spacing and fins per inch.

"The heat transfer points of a radiator are where temperature is actually allowed to leave the radiator and that occurs where the fins are bonded to the tubes. The more transfer points a radiator has the greater the temperature drop will be between the inlet and outlet. For comparison a '60s-style core typically had tubes spaced 1/2 inch apart, that is 1/2 inch of fin between the tubes. By going from a two-row radiator to a four-row core design we were able to double the heat transfer points, which resulted in a 15-20 percent increase in temperature drop without changing the other variables, such as air or coolant flow. U.S. Radiator offers four distinct core designs, the Standard that is found in most OEM-style radiators, the High Efficiency aluminum with 20 percent more heat transfer points, the High Efficiency copper/brass with 20 percent more heat transfer points, and the Optima copper/brass that uses 1/4-inch tube spacing with 1/4-inch fins that provides 40 percent more heat transfer points

"Radiator materials have created quite a bit of controversy. In the '80s the Japanese came out with a core design in response to the need to downsize radiators that has become the standard in the industry because it was efficient enough to allow the re-introduction of aluminum (a less efficient heat transfer material) at the OE level. By changing the tube spacing to 3/8 inch, a core design referred to as High Efficiency in the industry, more tubes or water passages and fins were allowed across the face of a core with a specific width in inches. The design was simple enough but proved to be very efficient in that more heat transfer points created greater temperature drop inlet to outlet. It should be pointed out that the move to aluminum radiator construction was purely financial. The pound purchases the raw materials necessary to build a radiator and a finished aluminum radiator weighs about 25 percent of a copper/brass unit (dollars per pound being almost equal at that time). The result was a huge financial savings for the car companies.

"When it comes to the difference in performance between copper/

To keep or Road Tour Chevy cool we chose a triple-pass radiator from U.S. Radiator. Rather than a single fan we opted for a pair mounted in a shroud, also from U.S. Radiator.

An important factor is to match the opening in the core support to the size of the radiator so airflow is not restricted.

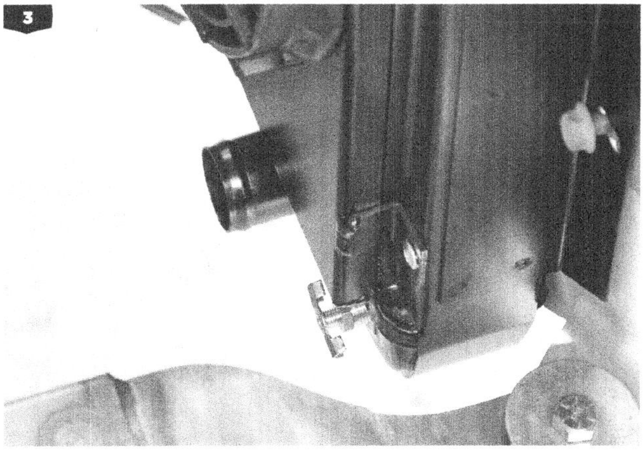

The custom fan shroud is securely fastened to the radiator's side straps. A brass petcock simplifies changing coolant.

The bung in the upper tank may be used for a fan switch or plugged.

The water pump and accessory drive system is a Front Runner from Vintage Air.

Due to the volume of radiators produced, large runs of tanks are made. Here is a supply of aluminum tanks in front, with copper versions in the rear.

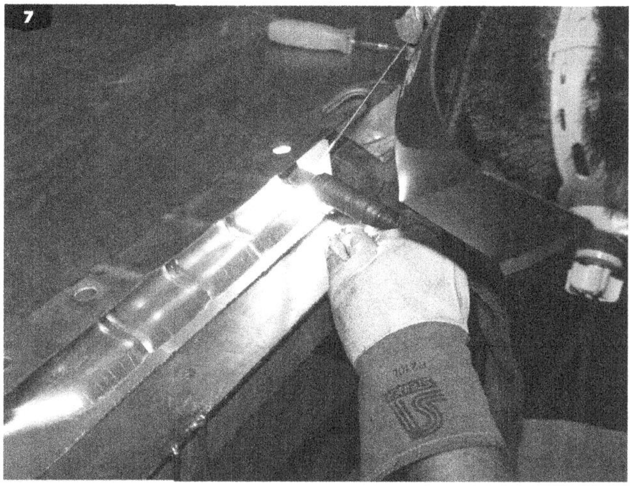

Here one of the baffles for the Triple-Flow option is being welded in an aluminum tank.

Hot Rods by Dean connected the radiator to the Shaver-built engine with stainless steel tubing and short sections of hose.

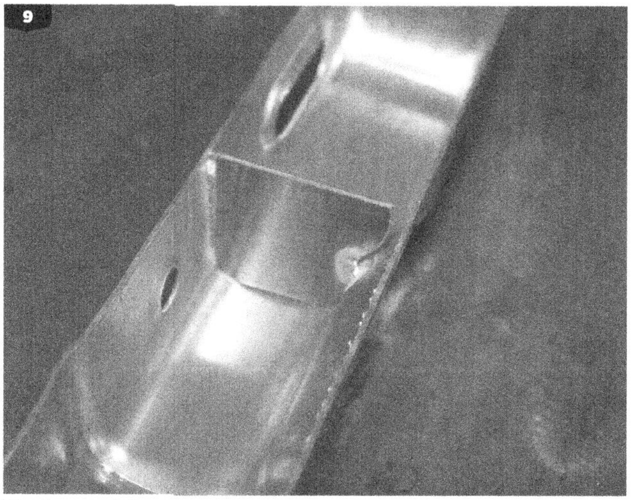

This is one of the completed directional baffles in a Triple Pass radiator.

The first step in assembling a copper/brass radiator is putting the core in a fixture and attaching the top header.

Next the bottom header is installed.

Once the headers are in place this roller is used to make sure the openings in the ends of all the tubes are uniform.

The core assembly is sent through an oven, which melts the solder on the tubes, securing them to the fins.

brass and aluminum radiators you may find the tests by U.S. Radiator surprising. They found that temperature drops at all operating ranges were virtually the same, with a slight advantage going to the copper/brass unit. But consider this; the thermal conductivity or heat transfer rate of copper is 92 percent versus aluminum at 49 percent. However, the copper fin is bonded to the tubes or water passages using lead solder, which is very inefficient and slows the heat transfer rate to just slightly better than that of aluminum. This can be a disadvantage if the bonding process does not allow the copper fin to touch the brass tube and why not all copper/brass cores of similar design but different manufactures transfer heat equally. Copper/brass radiators, because of their weight and durability, have been around a long time and are easily disassembled and reassembled for cleaning purposes. Not the case with aluminum unless speaking of the OE version that comes with crimp mounted plastic tanks. As a result the life expectancy of the aftermarket aluminum radiators will be far less than that of copper/brass." ★

14

After the cores have cooled sufficiently the headers are dipped in solder to permanently attach the tubes.

15

Here is a header that has been soldered. Note this radiator has four rows of tubes.

16

These completed cores will have been matched with tanks with the proper inlet and outlet locations.

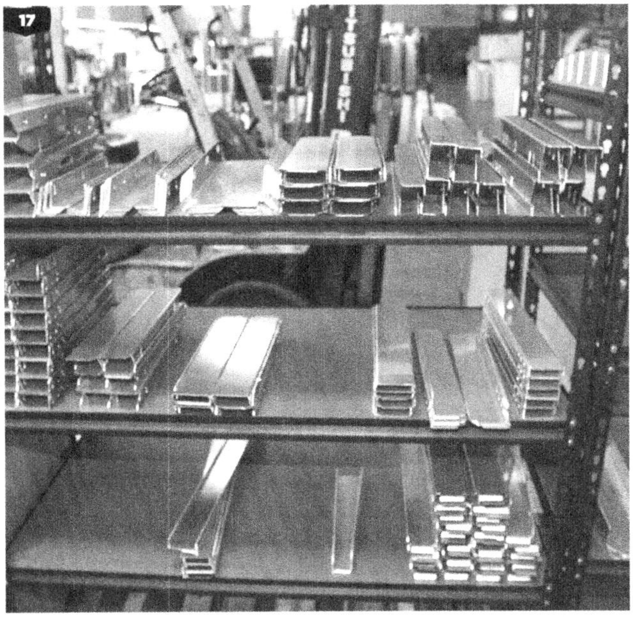

These are side straps, or mounting brackets, for aluminum radiators.

A Thermo-Anemometer is an evaluation tool that can be used to measure airflow or air temp.

U.S. Radiator can supply custom fan shrouds for virtually any their radiators; they can be used with mechanical fans or mount an electric.

Don Armstrong uses it for both to evaluate cooling systems. Here he checks the airflow of an electric fan and shroud installation.

The Triple-Flow Option

In most radiators coolant flows from one end of the core to the other, so it stands to reason that the longer the core is the more exposure the coolant has to airflow, which would result in better cooling. Basically that's the reason cross-flow radiators were developed. As cars became lower, the only way to provide adequate core length was to make radiators shorter and wider and run the coolant side to side.

Of course the problem with street rods is that the length of the core is restricted by the space available. To address that issue U.S. Radiator developed the Triple Pass radiator. This unique design uses a series of baffles to re-route the coolant through the radiator core three times. As a result the coolant travels further through the core and coolant temperature can be dropped 15 to 20 percent. The Triple-Flow option is available in down-flow and cross-flow designs.

CHAPTER SEVEN

Bulletproof

Building a Transmission and Driveshaft

By Ron Ceridono | Photography by Brian Brennan and Courtesy of Dynotech

When automatic transmissions first began to show up in street rods they received mixed reviews—there were those who found the ease of installation to be an advantage while others were suspicious of their performance potential. Since those early days of four-speed Hydramatics and two-speed Powerglides, automatic transmissions have continued to improve, and specialists like Hughes Transmissions can make them even better.

This is the completed Hughes Extreme Duty 4L80E transmission assembly for our Impala. From left to right: Tony Kane (Hughes' general manager), Dean Livermore (owner of Hot Rods By Dean, builders of the 2014 AMSOIL/*STREET RODDER* Road Tour '59 Chevy), and Danny Gutierrez (Hughes transmission technician).

Each Hughes transmission is completely dyno tested to verify line pressure, shift function, and cooler flow before it goes out the door.

Hughes has an array of heavy-duty, high-performance parts for the 4L80E. Here is a comparison of stock GM components at the top—34-element intermediate sprag, intermediate outer sprag race, and direct drum. Lower components are Hughes Extreme Duty parts: 36-element intermediate sprag, Extreme Duty heat treated extra wide intermediate outer sprag race (allows use of up to five intermediate frictions and steels instead of three), and Hughes-modified GM direct drum.

On the left is a GM direct drum modified by Hughes to accept 36-element intermediate sprag, right is OEM GM direct drum for a 34-element intermediate sprag. Sprags are one-way clutches and the stockers are weak points when lots of horsepower is put to them.

On the left is Hughes Extreme Duty 36-element intermediate sprag that fits the wider, modified race, right is stock GM 34-element intermediate sprag.

One of the most popular of the new breed of automatic transmissions is General Motors' 4L80E, and that's what we chose to get the AMSOIL/*STREET RODDER* Road Tour '59 Chevy into gear. The Hughes Transmission crew has years of experience with this transmission and they feel it is the perfect choice for those who want the strongest overdrive automatic available and that was reason enough for our decision.

According to Hughes there is much about the 4L80E. They have a gear spread that allows hard, off-the-line acceleration with an overdrive top cog for reduced engine rpm for long distance cruising—ratios are First, 2.48:1; Second, 1.48:1; Third, 1.00:1; and Fourth, 0.75:1. In addition these transmissions can be made to withstand huge amounts of horsepower with modifications Hughes has perfected. Add to that a torque converter that allows higher-than-stock stall speed for maximum torque multiplication along with a lock-up function, which virtually eliminates all heat buildup in the converter at cruising speeds and improves efficiency, and you've got the best in terms of performance and practicality.

Like all contemporary automatic overdrive transmissions the 4L80E requires an electronic control unit to operate. Hughes offers controllers that are pre-programmed and require no software, laptop, or PC for installation, tuning, or proper function and a pre-terminated wiring harness for simple plug-and-play operation is included. When required, a throttle position sensor (TPS) kit and mounting bracket that will work with virtually any carburetor is optional, with EFI systems the Hughes harness is simply spliced into the existing TPS.

Automatic Transmission Basics

While automotive automatic transmissions may have two, three, four, or more gears, manual or computer controls, and vary in complexity, the basic internal parts remain the same, there are just more of those parts in some transmissions than others. Here's a look at the major components of an automatic transmission and some of the usual performance upgrades:

Torque Converter

This is what connects the engine to the transmission. Not only does it allow the engine to keep running while the car is stationary, but it can also provide torque multiplication when accelerating from a standstill. For a simplistic explanation of how a torque converter works, picture what would happen if two electric fans were facing each other and one was turned on. As the one under power began to turn and move air, the other one would also begin to turn. If you can visualize that, you've got the basic idea of how a torque converter works. In a torque converter, both fans are in a container: one is connected to the engine and the other to the transmission, and oil is used rather than air.

As might be expected, there is some slippage inherent with a torque converter; so many overdrive automatics now have converters

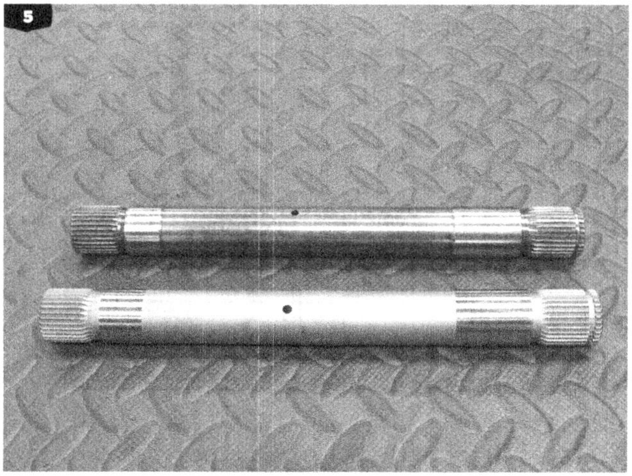

Hughes billet steel intermediate shafts (bottom) are considerably stronger than the OEM part.

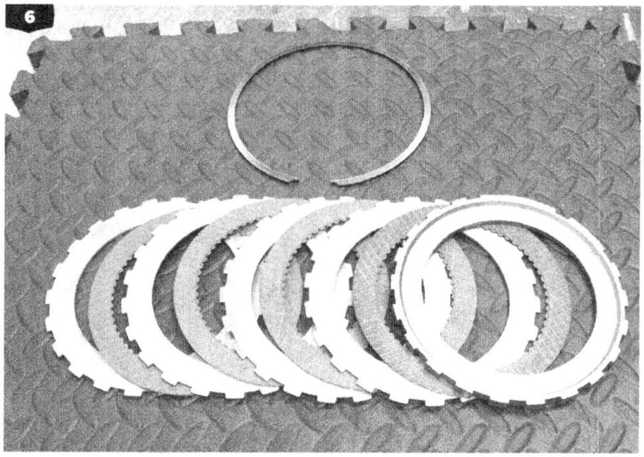

Clutch packs must be able to transfer power without slipping. Hughes' intermediate clutch pack (those responsible for Second gear) is beefed up with a billet steel pressure plate, five High Energy intermediate frictions, five intermediate steels, and a heavy-duty snap ring.

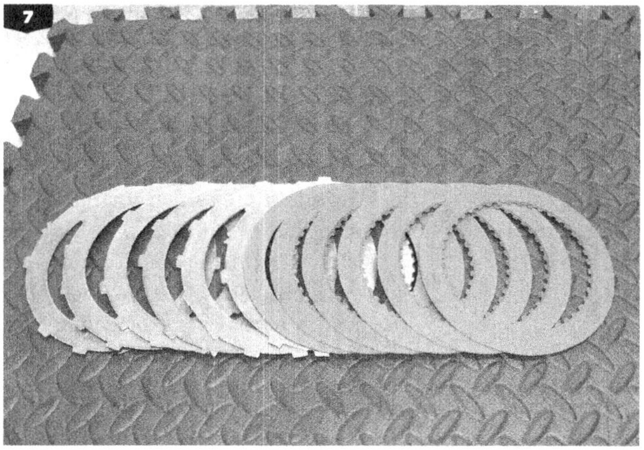

The direct clutch packs (responsible for direct drive or Third gear) are also improved with six red Alto direct frictions and six direct steels.

On the left is a forged steel forward clutch hub with Torrington roller thrust bearing for reduced friction and resistance to wear, on the right is the OEM GM forward clutch hub with thrust washer.

with a hydraulically applied internal clutch that hooks the transmission directly to the engine for increased efficiency. While better fuel mileage is often considered to be the advantage of a lockup converter there is another, often overlooked, purpose. The higher ratio provided by an overdrive Fourth gear can put an additional load on a conventional converter causing excessive slippage even in light throttle cruise conditions. That slippage creates heat and heat's the enemy of an automatic transmission. When used with an overdrive automatic a lockup style will lower engine speed in cruise conditions and lengthen the transmission's life. Overdrive automatics and lockup converters are made for one another.

Planetary Gears

As gears go, planetaries can do it all. Made up of three elements— sun, ring, and planet pinion gears—they can provide forward or reverse rotation, a speed increase, constant speed, or a speed reduction.

Three things are necessary to make planetary gears operate: an input (power from the engine), an output (power going out), and a reactor (one of the elements is held stationary). The gear ratio and the direction of travel depend on which element is performing each function. Most transmissions have more than one planetary gear set to provide a variety of gear ratios.

Bands And Clutches

Bands and clutches hold the reactors stationary. An automatic transmission goes into gear by holding one part of a planetary gear set stationary with a band or a clutch that is applied by hydraulic pressure. One reactor is released and another applied when the transmission shifts gears; the transmission is in Neutral

9

Here is a Hughes-modified reaction carrier with Torrington roller thrust bearing on the left, right is OEM GM reaction carrier with thrust washer. Again the purpose is reducing friction and wear.

10

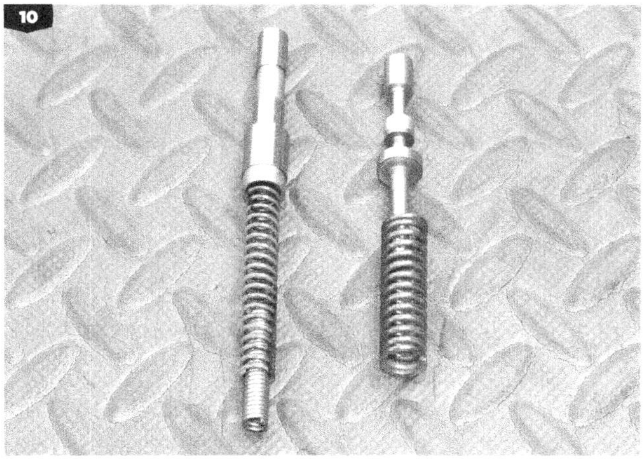

Hughes makes improvement in the shift body with new valves and springs.

11

Improved bands are part of the Hughes' package, that's a Kevlar intermediate band on the left; right is high static rear band.

12

Compare the narrow OEM transmission case bushing, to the wider version. The shouldered design of the aftermarket bushing that prevents it from walking out of case.

if no reactor is applied. If you've ever been in a car and the transmission felt like it was slipping, that's exactly what was happening—the band or clutch pack wasn't holding the reactor stationary and it was slipping and not transferring full power. In an extreme case, the clutches, or band, don't hold at all, and one or more gears (and in some cases, every gear) stops working as a result.

Valvebody

The valvebody is the hydraulic "brain" of the transmission; it controls the shifting of gears by controlling which reactor is applied and when. Some transmissions use hydraulic pressure from a governor, throttle valve, or vacuum modulator to determine shift points, while many contemporary versions use computer-controlled electromechanical servos.

Transmission Modification

Let's start with the torque converter. In many cases, the stall speed of the converter is increased. Simply put, stall speed is the rpm that the engine will reach with the transmission in gear, the brakes applied, and the throttle held wide open. The higher rpm simply allows the engine to produce more power, which will launch the car harder from a standstill. The downside is that higher stall speed converters will slip more in "normal" use. That can create excessive heat, which is what damages transmissions. Of course the perfect solution is a lockup converter with increased stall speed—that's the best of both worlds.

Shift Kit

Most stock automatic are designed to be seamless; that is, the shifts are smooth to the point of being hard to detect; this is done by timing the release and application of the various reactors. One may begin to apply as

To ensure true and flat sealing surface, prevent cross leaks between pump stator and pump body, and eliminate cavitation of the pump gear set and/or torque converter, Hughes machines the mating surfaces.

Show here are *(upper left)* the pump stator, *(upper right)* the pump body, *(lower left)* outer pump gear, *(lower right)* and outer inner pump gear. These are the four components of transmission's pump.

The cleaned, refurbished, and painted transmission case is ready to receive its internal components.

These parts are the direct clutch hub (left) and the center support *(right)*.

the other begins to release so there is a split second of overlap. While this results in a smooth shift, there is a certain amount of slippage that takes place in the process, which wears the friction surfaces.

Shift kits generally do two things: They change the timing of the release and application of the reactors, which results in a firmer, more noticeable shift and may also lengthen the life of the transmission. In addition the hydraulic pressure to apply the reactors is often increased too, which means the clutches and bands have more holding power, thus increasing the torque capacity of the transmission.

With many transmissions today using computer controls, a variation on the shift kit theme is the hopped-up computer. It can and will do the same thing as a conventional shift kit, but changes can often be made without dropping the trans pan.

Friction Surfaces

A common method to increase an automatic transmission's torque-carrying capacity is to improve the friction surfaces that hold the reactors. This may be done by using improved materials, more clutch plates, wider bands, or a combination of them all.

Hard Parts

Just as with a manual transmission, an automatic has mechanical components that are susceptible to damage when overstressed. As the horsepower applied to it is increased, it is often necessary to increase the strength of an automatic transmission's shafts, clutch drums, planetary gear sets, sprags, and other internal parts.

Selecting A Transmission

Hughes Performance has a 4L80E transmission available for practically any engine combination and power level. Their Street/Strip series is

Here Danny installs the gear train into transmission case. The gear train includes the intermediate shaft, sun shaft, center support, reaction carrier, output carrier, ring gear, and output shaft.

The rear servo is bolted in place; it applies rear band to help create reverse gear.

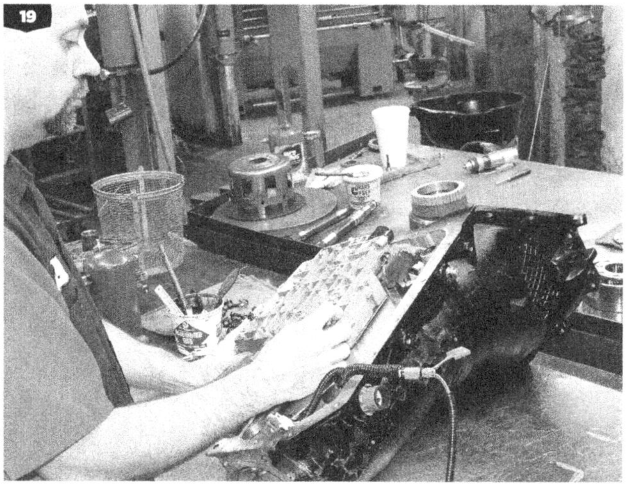

Next comes the installation of upper valvebody, internal wiring harness, and the case connector.

a great choice for combinations producing up to 850 flywheel horsepower; the Heavy Duty Street/Strip series 4L80E that is capable of withstanding 1,000-plus flywheel horsepower and for those running then there is the Extreme Duty Street/Strip series 4L80E that is rated for use with 1,500-plus flywheel horsepower. Keep in mind, the object of selecting a transmission is not to impress your friends but get the right transmission for the application. A crate motor won't require an Extreme Duty transmission, but if the engine in question is making 603 lb-ft of torque and 621 hp, like our Ron Shaver–built 430-inch LS, some improved parts may be necessary. The best way to get the correct transmission for your application is to do what we did with the AMSOIL/*STREET RODDER* '59 Chevy—give Hughes a call and let them make a recommendation.

Valvebodies

When it comes to selecting a valvebody there are a number of choices.

The 4L80E features electronic solenoids built into the valvebody. These solenoids control the shift functions, regulate the line pressure generated by the front pump, and regulate the torque converter lockup clutch function. Many modern aftermarket EFI controllers will control this transmission as well, which means a separate TCU would not be required.

Another option is a full manual valvebody. It features a forward shift pattern (P-R-N-4-3-2-1), and requires the driver to manually upshift and downshift the transmission. This valvebody option will not shift automatically, and eliminates the necessity of an electronic transmission control unit. The torque converter lockup function

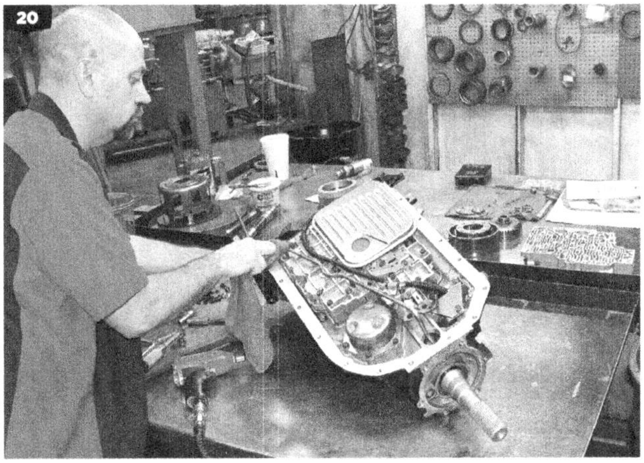

20 The last internal parts to go in place are valvebody and filter assembly.

21 The last step before going on the dyno is the installation of deep cast aluminum transmission pan. It's equipped with a drain plug for easy servicing and holds 4 extra quarts of fluid for additional cooling capacity.

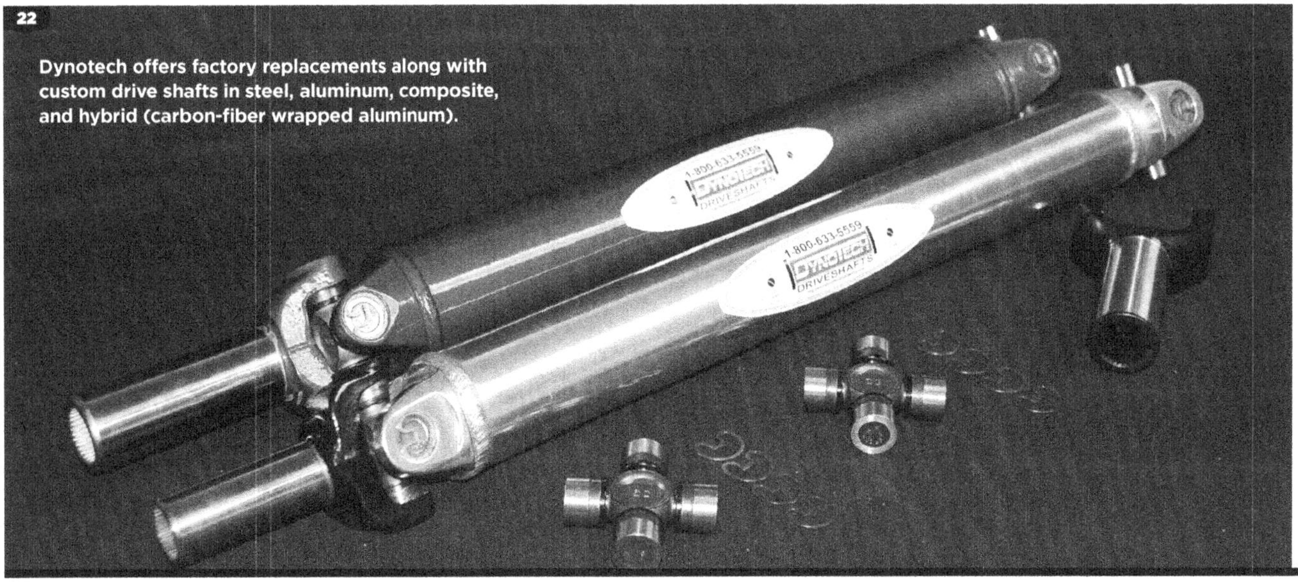

22 Dynotech offers factory replacements along with custom drive shafts in steel, aluminum, composite, and hybrid (carbon-fiber wrapped aluminum).

can still be retained and operated via a simple toggle switch or push button that will have to be manually activated by the driver. The full manual valvebody option is a great choice for the enthusiast who wants to save a few dollars on a 4L80E package and does not mind shifting the transmission manually.

The third valvebody option is a full manual transbrake valvebody. It features a forward shift pattern (P-R-N-4-3-2-1) and requires the driver to manually upshift and downshift the transmission. The transbrake feature allows the transmission to essentially be "locked up" in low gear so that wide-open throttle can be applied without experiencing any vehicle movement. When the transbrake is released via an electric button the transmission unlocks and allows the vehicle to launch under maximum power. This option is very popular for street/strip enthusiasts who want the benefits of an overdrive transmission on the street, maximum performance potential at the dragstrip while eliminating the necessity of a TCU.

Our fourth valvebody option is an automatic transbrake valvebody. This valvebody option requires the use

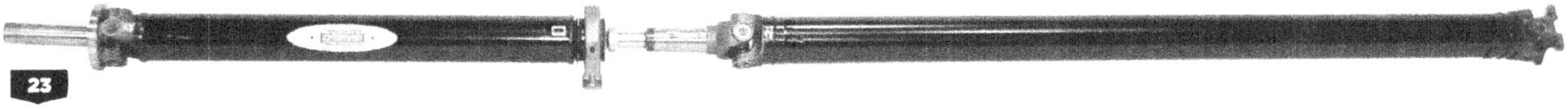

23

For applications that require them (like a stock '59 Chevy) Dynotech offers two-piece driveshafts.

24

Also available from Dynotech are center supports for two-piece 'shafts.

25

The first step to driveshaft construction is cutting the tubing to length.

26

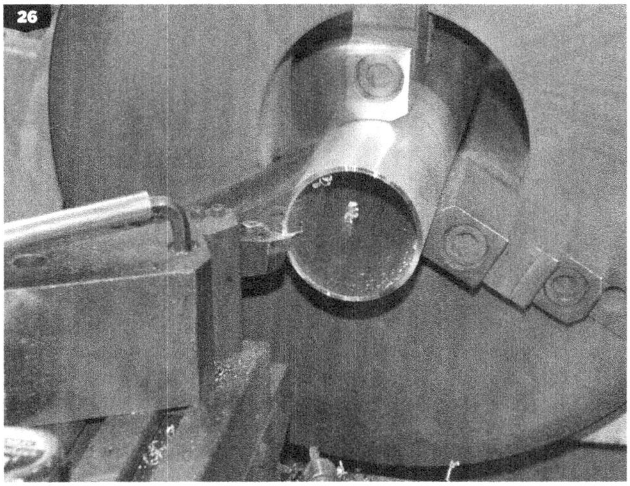

Next the cut ends are made perfectly square in a lathe.

27

The yokes are pressed into the tube while overall length is checked.

of a special electronic transmission control unit. This very unique valvebody option allows the driver to place the shifter in "drive"(Third gear), activate the transmission, launch the car, and the transmission will then automatically upshift based on the settings programmed into the electronic transmission control unit. This valvebody option is ideal for the enthusiast who wants the comfort and ease of automatic shifting, along with maximum performance potential at the dragstrip.

Torque Converter Selection

When designing a torque converter Hughes gathers as much data about the customer's combination as possible, such as crankshaft stroke length, camshaft duration, camshaft lobe separation angle, torque production, vehicle weight, rear gear ratio, and rear tire height. Use of poweradders will also influence how the converter will be built.

Hughes offers 9-, 10.5-, and 13-inch torque converters for our 4L80E transmission packages, with stall speeds ranging anywhere from 1,800 to 5,500-plus rpm, depending on what the combination requires. The majority of Hughes 4L80E converters feature a billet steel front cover and billet steel lockup piston with extra-wide heavy-duty clutch lining for maximum durability. They also offer triple lockup disc options for those customers wanting the absolute strongest torque converter that money can buy.

Coolers

When it comes to transmission coolers, it is very difficult to overcool an automatic transmission. Hughes recommends that the customer install the largest external transmission cooler they can fit in the car and that fits their budget.

Building A Driveshaft

Dynotech, formally known as Balance Engineering, built the custom driveshaft connecting the Hughes transmission to the Roadster Shop IRS centersection in the Road Tour '59. Dynotech Engineering Services specializes in manufacturing custom-made driveshafts that are hand assembled, computer welded, and high-speed balanced and what's really cool is they are as close to you as the nearest telephone. Supply Dynotech with the necessary information and incredibly in most cases a driveshaft will be delivered to your door within 72 hours—and sometimes less.

Dynotech's driveshafts are guaranteed vibration-free, and they stand by their work 100 percent. All of their driveshafts are assembled by hand while using modern technologies, such as computer welding and high-speed balancing. As a result Dynotech's driveshafts are held to remarkably close tolerances.

Driveshaft Materials

Dynotech offers driveshafts in steel, aluminum, composite, and hybrid materials:

28 At each end of the press are fixtures that ensure the yokes are in perfect alignment.

29 New U-joints of the proper series and a new slip yoke are included in complete driveshaft assemblies.

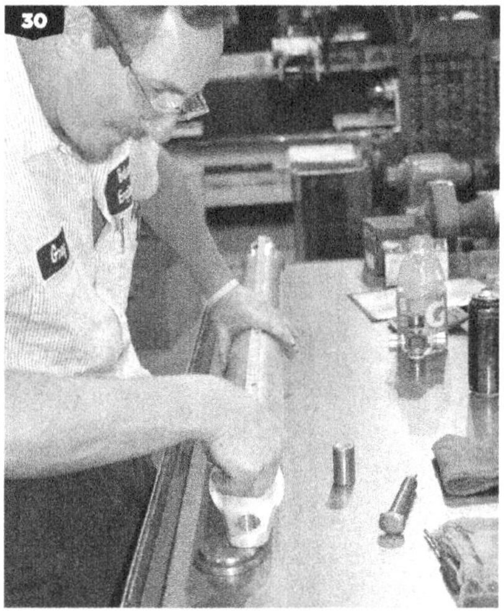

30 After the yokes are pressed in place overall length is checked again. All driveshafts are built to within 1/16-inch (0.0625-inch) in length.

31 Dynotech assembles all of its driveshafts by hand while using modern technologies, such as computer welding and high-speed balancing. As a result Dynotech's driveshafts are held to remarkably close tolerances.

Steel: Made from DOM (drawn over mandrel) seamless chromoly tubing, these driveshafts are for high-performance street and racing applications. They have higher torque and rpm ratings than OEM shafts and come in 2-, 2.5-, 3-, 3.5-, and 4-inch diameters with 0.065, 0.83-, and 0.95-wall thickness. Chromoly shafts are for serious racing applications, with increased torsional strength designed to resist well over 2,000 hp. But while they are super strong, a chromoly shaft is significantly heavier than an aluminum shaft of the same size.

Aluminum: These 6061-T6 aluminum shafts are for performance, street, racing, and stock replacement applications. They are lightweight and come in 3-, 3.5-, and 4-inch diameters with 0.125-inch wall thickness. Yokes accommodate 1310-, 1330-, and 1350-series U-joints. Dynotech's aluminum driveshafts can comfortably take up to 1,000 hp and around 900 lb-ft of torque. For 99 percent of performance and race cars out there, the weight savings and high horsepower rating of an aluminum driveshaft would be enough.

Carbon Fiber: Intended for extreme high-performance and racing applications, these shafts are strong, resistant to flex, and they are expensive. They are available in 3.5- and 4-inch diameters with steel or aluminum yokes for series 1310, 1330, 1350 and 1410 U-joints

Hybrid: Carbon-fiber wrapped aluminum driveshafts are for high-performance racing applications. These shafts carry the highest rpm rating and come in 2.75-, 3-, 3.5-inch diameters with 0.125-inch wall thickness. Yokes for 1330 and 1350 U-joints are available. *

32

Here a new U-joint is installed in the pinion shaft end of the driveshaft.

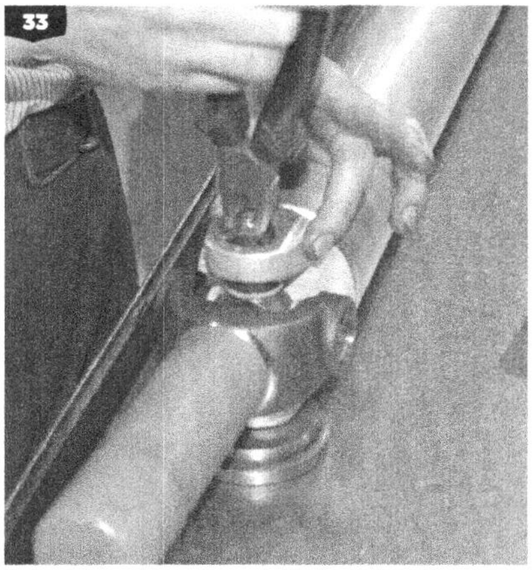
33

In the transmission end a new U-joint and slip joint are installed.

34

The assembled driveshaft is placed in the automatic welder. Runout is checked during welding and is held to less than 0.010-inch.

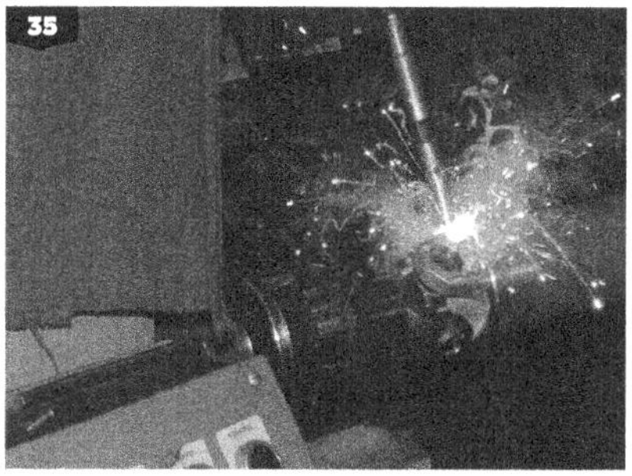
35

As the driveshaft turns 360 degrees the automated welder runs a continuous bead.

36

The finished weld is perfect.

37

After assembly and welding all driveshafts are checked for runout then high speed balanced between 5,000 and 8,000 rpm to less than 0.20 ounce/inch.

CHAPTER EIGHT

Good All Under

A New Floor

By Ron Ceridono | Photography by Brian Brennan

Riding on a Roadster Shop chassis and full of parts from our official supplier, Speedway Motors, this Chevy takes us further into the '50s than we've ventured in the past. Nonetheless, this is a street rod through and through with the horsepower, handling, fit, finish, and attention to detail befitting the 19th edition of our Road Tour program.

Given the performance expectations for this car we quickly came to the conclusion the stock GM chassis wasn't going to cut it. The cruciform design, often referred to

Here's a sneak peak of what's to come, the AMSOIL/*STREET RODDER* Road Tour '59 Chevy body in Ferrari Red. It's ready to drop on the new Roadster Shop Fast Track perimeter frame.

as an X-frame, lacked the torsional rigidity to meet our needs. As the saying goes, adding performance suspension to the stock 'rails would be like putting a $100 saddle on a $10 horse. The solution to this dilemma came in the form of a Roadster Shop '58-64 Fast Track perimeter frame with custom independent front and rear suspension.

Once the new chassis was delivered to Hot Rods by Dean the process of combining old and new began. There was rust damage to repair (primarily in the trunk) and due to the differences in the old and new frames, some modification of the Impala's floor were in order, but given the advantages of the replacement chassis they were well worth the effort. The majority of the modifications necessary were below the front seat, in the rear seat footwell, and in the trunk. Of course some of the modifications, notably the rear wheelhouses, were due to the massive Michelin tires.

With the floors repaired and revamped all the welded joints were given a coat of Eastwood seam sealer. The bottom of the floor was protected with Eastwood's rubberized undercoating, while the topside was given the vintage treatment with spatter paint (several color combinations are available from Eastwood in spray cans) before most of the interior surfaces were covered with Dynamat. ⋆

The stock Chevy frame lacked the torsional rigidity to cope with the high-performance suspension we wanted.

Our Roadster Shop chassis is everything the original wasn't—strong, stiff, equipped with high-performance suspension at both ends, and it's great looking to boot.

To accommodate the new chassis it was necessary to modify the passenger compartment floor.

Viewed through the rear window the amount the 'rails protrude below the passenger seat can be seen.

Other than trimming to clear the new upper control arms and coilovers the front wheelhouses were not modified.

The rear wheelhouses were cut away to be widened for the 285/35ZR-19 rubber.

Modifications to the trunk floor were necessary due to shape of the Roadster Shop frame and Fast Track IRS.

The original trunk floor was rusty and would have been replaced. The floor between the wheelhouses would have to be narrowed to fit between the new wheel tubs.

Here the rear floor section has been removed. Note the wheelhouse has also been split lengthwise.

The rear wheelhouses were completely removed; the factory spot welds were removed with an Eastwood spot weld cutter.

Mini-tubs were created by adding strips of sheetmetal to the middle of the stock wheelhouses.

Widening the wheelhouses meant pockets had to be built in to accommodate the trunk hinges.

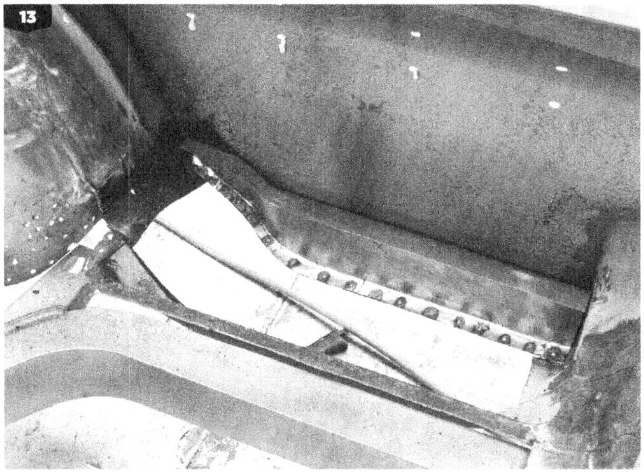

Part of Hot Rod by Deans trunk rebuild included fabricating new inner fender sections and floor supports.

This is the stock trunk floor, the centersection was rusty. Note the layout lines indicating where it will be cut.

Seen from the bottom old floor viewed from the bottom. The gas tank reinforcements will be recycled; the proper distance between the supports was preserved by tack welding lengths of tubing to them.

The salvaged gas tank supports will be attached to the new trunk floor.

17

Caged nuts would be used to secure the gas tank straps to the support.

18

Once the locations of the gas tank straps were determined holes were cut in the floor and the caged nuts were put in place.

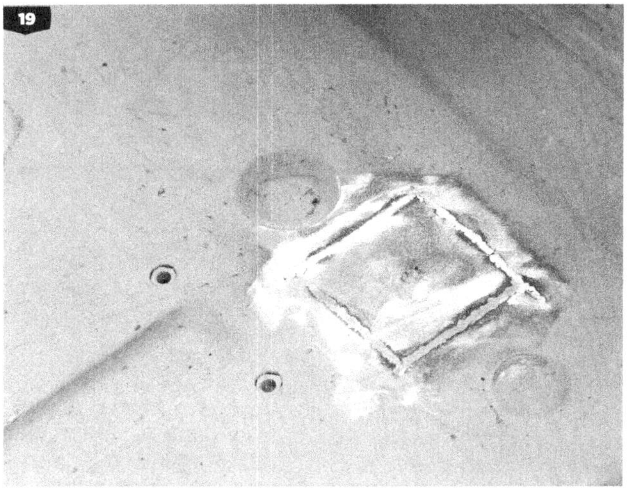

19

With the nuts installed the holes in the floor were plugged.

20

This is a worm's eye view of the new trunk floor and the supports the gas tank will attach to.

21

Hot Rods by Dean gave their new Eastwood English wheel a workout forming new tunnels for the transmission and driveshaft.

22

The transmission tunnel was fabricated in multiple sections and then welded together.

Likewise the driveshaft tunnel was made in two pieces then joined with a Miller welder. The new tunnel is taller than the original due to the car's lower stance.

Sheetmetal "boxes" were made and welded in place to enclose the floor in those areas where the new frame encroached into the passenger compartment.

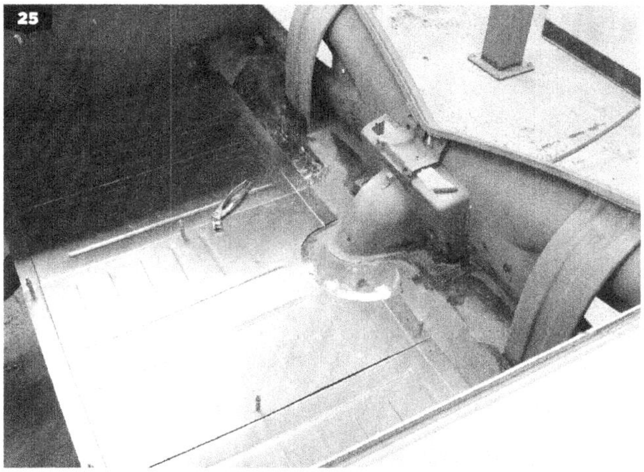

The replacement floor panel was held in place with Eastwood clecos while rossete welds were used to secure it.

In the passenger compartment the floor was given a coat of primer, then all the joints were covered with Eastwood seam sealer.

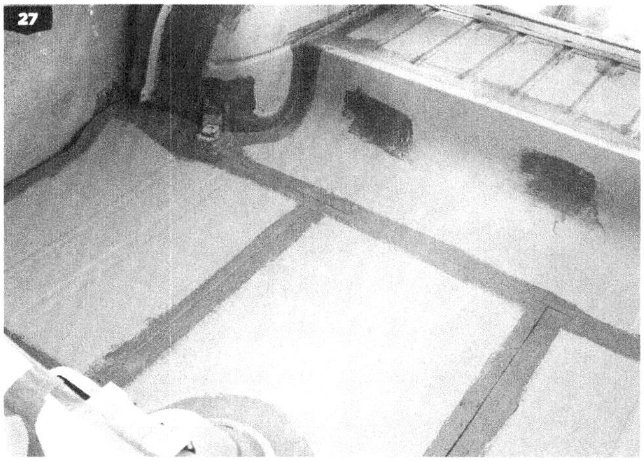

The trunk was given the same primer/seam sealer treatment. Seam sealer is highly recommended to keep deadly exhaust gas out of the interior spaces.

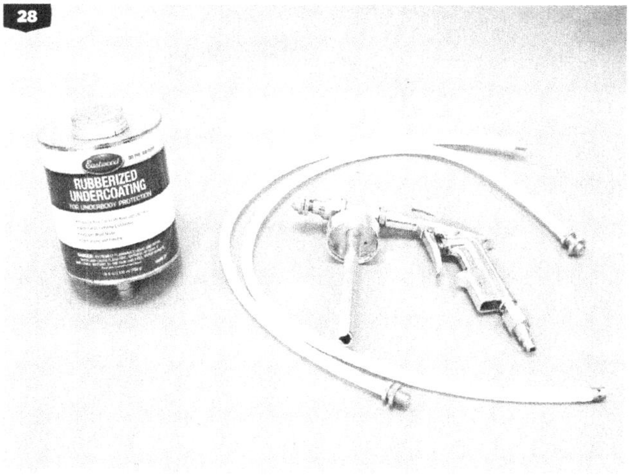

Eastwood's undercoating gun can be used to heavy-duty antirust, rubberized undercoating and spray-on bedliners.

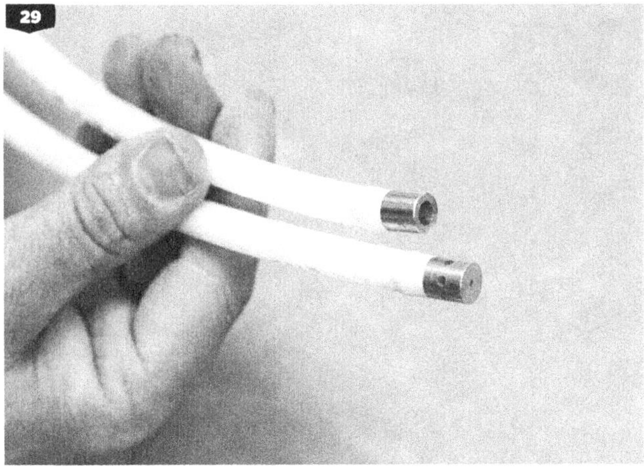

For getting into hard-to-reach areas, like rocker panels, Eastwood supplies two 24-inch flexible wands—one has a straight spray tip, the other a 90-degree.

Eastwood's rubberized undercoating stands up to the most extreme conditions. The tough, flexible film prevents rust and corrosion and it may be painted.

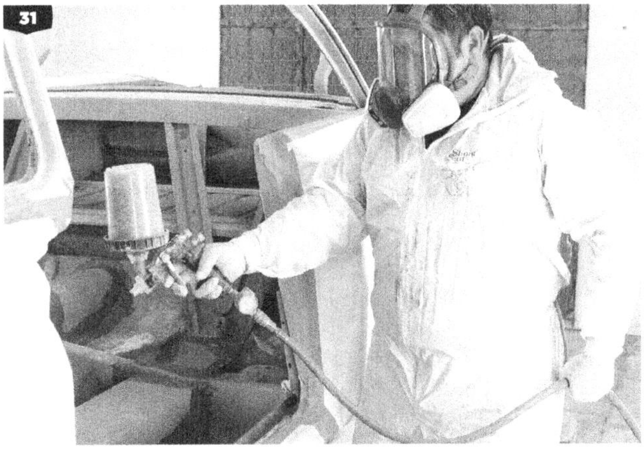

With the floor modifications and repairs made, and the seams sealed all the surfaces were primed and then covered with vintage-looking spatter paint.

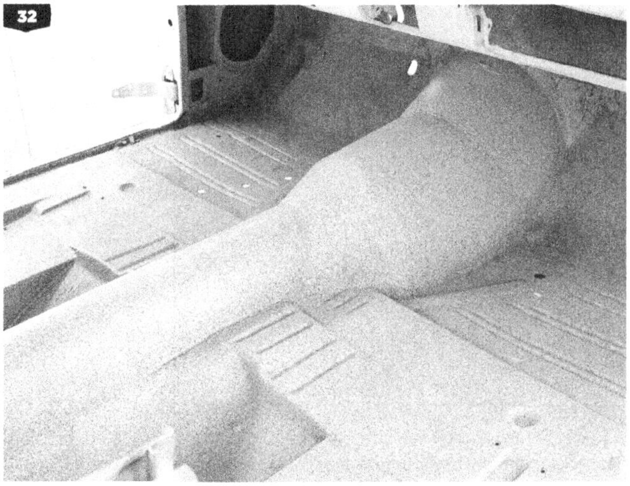

To most observers the floor appears original.

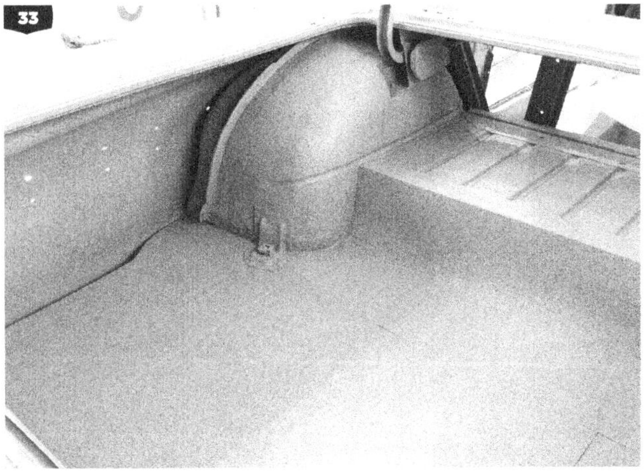

With the seam sealer and spatter paint even the modifications to the wheelhouses are hard to detect.

Like Road Tour cars of the past, the interior surfaces have been layered with Dynaliner and Dynamat to reduce noise and heat in the passenger compartment.

35 To prevent heat and noise absorption the underside of the roof was treated to a layer of Dynaliner.

36 All the rubber products, including the body mounts came from Soffseal.

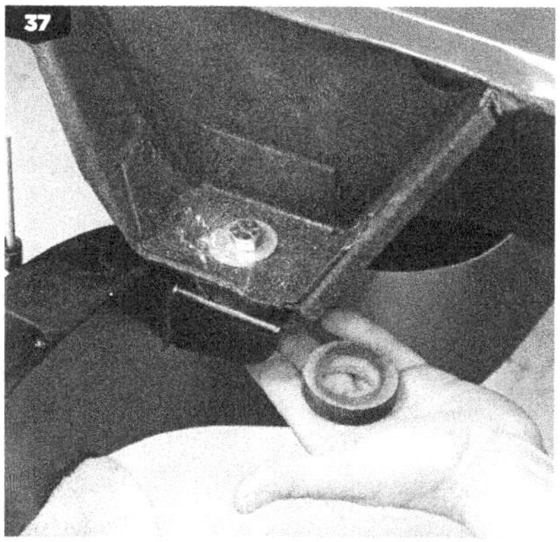

37 The body mounts register in holes in the frame. The lip on the bushing is to locate the lower portion of the mount; the metal sleeve prevents the mounting bolts from being over tightened and crushing the rubber.

38 It took the whole crew to put the body back on the frame—there was considerate grunting and groaning.

CHAPTER NINE

Finishing Touches

Ferrari Red Paint, Polished Aluminum, and Lots of Chrome

By Ron Ceridono | Photography by Brian Brennan

While our '59 Chevrolet came to us in reasonably good condition for a car of its age, it had some rust damage, an assortment of lumps and bumps that had accumulated over the years—and the truth is these cars were surprisingly wavy from the factory. It was going to take hard work and knowhow to get the Impala to Road Tour standards, but the team at Hot Rods by Dean was up to the challenge.

Basically the processes used by Dean and his crew are standard body shop procedures, however what will set this car apart is the level of workmanship. Every square inch of the Chevy's sheetmetal

Looking like it was dipped in lipstick, Hot Rods by Dean has our AMSOIL/*STREET RODDER* '59 Chevrolet closer to being Road Tour ready.

This view gives one a sense of how big a '59 Chevy is. Our '59 Impala was reasonably solid and was sans engine, trans, and some big pieces of glass.

was stripped by a combination of sanding and media blasting. Once that was done the corrosion repair that is part of reviving any vehicle of this age was carried out the only way repairs are permanent—the rust was cut out and new metal was welded in place. Thanks to a shop full of Eastwood equipment, all the necessary replacement panels could be produced in-house regardless of their complexity.

With what was now a rust-free body the next chore was to make the acres of sheetmetal perfectly smooth. Careful bumping with hammers, dollies, spoons, and all the other tools of the trade and a trace of filler here and there and the body is nice enough for an average build—but Livermore doesn't understand the concept of average. An application of PPG's VP2100 Primer/Filler, followed by block sanding, then more VP2100, followed by more block sanding, provided the desired results: perfect '59 Chevy sheetmetal.

After a final application of PPG's K38 primer the Impala was ready to be covered in PPG's DBC 908276W—or as it's better known, Ferrari Red. A coat of PPG's D894 high-solids clear, a round of color sanding and polishing, and the results speak for themselves.

POINTERS ON PICKING PAINT

Painting cars has changed since our Impala was new. Back then acrylic enamel and acrylic lacquer were popular and each had unique characteristics. Enamel was tougher, but was far more difficult to apply. It was not possible to color sand and polish enamel, so what came out of the paint booth was what you lived with. Lacquer was certainly easier to work with but it required polishing, which took time but the results were worth the effort. As a result of these characteristics most repair shops chose enamel, while custom painters continued to use lacquer.

An interesting development came with the advent of hardeners for acrylic enamel. Appearing to have the best attributes of both lacquer and enamel, many painters believe that if a little hardener was good, a bunch would be great. The end result was often crazing and paint that came off in chunks.

Another gigantic improvement in the appearance and durability of automotive paints came in the late '70s with the introduction of two-stage, or basecoat/clearcoat finishes. At first expense was an issue, but it didn't take long for automobile manufacturers and repair shops to recognize the benefits of these better looking, higher gloss, more durable finishes.

While paints continued to improve they also became more toxic. Better personal protection gear was needed, improved booths, and the federal government got more involved because of something called Volatile Organic Compounds. VOCs are unstable, carbon-containing compounds that are released when solvent-based paints are sprayed as they dry. When VOCs enter the air, they react with other elements to produce ozone, which causes air pollution and a host of health issues. One method of reducing VOC's was with the introduction of a High Volume Low Pressure (HVLP) spray gun. More material makes it to the surface and less is lost to the atmosphere during application. Of course the real change came with how materials are formulated. VOC emissions are established by pounds per gallon of material. And while there are federal regulations that apply, VOC limits vary from state to state.

There is a long list of products used to produce a quality paintjob. Here is a look at the common materials used:

Primer: The function of any primer is to provide a surface for paint to adhere to. Primers by nature are thin and do little to fill imperfections. There is no advantage to heavy coats of primer; in fact it should be avoided.

Primer-Surfacers: These products are a foundation for the paint, plus they fill imperfections. Multiple coats of primer-surfacer will build a thick film quickly and allow for block sanding to get rid of surface irregularities and provide a super-straight base for the topcoat. Always keep this in mind: If the car is not straight in primer, it won't get any better with paint. That's the reason bodies are often block-sanded several times.

Contemporary primer-surfacers, such as those available from PPG, are catalyzed, which eliminates shrinkage and the sand scratches that used to result from the solvent evaporating from lacquer-based products. Additionally, primer-surfacers provide excellent corrosion resistance and a chip-resistance base for color coats.

Sealers: These products are

Vibrance Collection

The Vibrance Collection from PPG offers the latest in custom colors using unique micas and innovative pigments to create a truly custom finish. Ditzler Big Flake, Liquid Crystal are designed to create a three-dimensional effects. Harlequin's six color-shifting pigments can create colors that change completely when viewed from different angles. And of course there are pearls, candies, and "standard" metalflake. Luminescence II can be mixed with Deltron DBC basecoat to create a color travel effect on basecoat and tri-coat finishes and Flamboyance offers six custom shades that can be formulated to produce one-of-a-kind tri-coat effects.

Along with the basecoat products, there are Vibrance clears. Designed for custom applications, VC5225 Soft Touch Clear is high-solids, two-component urethane topcoat that provides a very low sheen for trim accents and interiors. VC5200 Custom Clear is a high-solid, urethane clearcoat with superior gloss.

1

The crew at Hot Rods by Dean began stripping paint off the large, flat portions of the body with a dual-action sander, the remainder of the Chevy's curvaceous sheetmetal was media blasted.

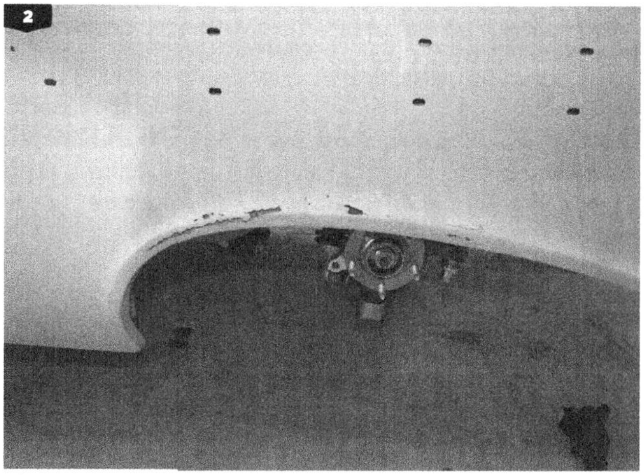

2

After stripping typical rust damage was discovered at the tops of the rear fender openings.

intended to keep the solvents in the topcoat materials from penetrating whatever is underneath as well as preventing undercoats from bleeding through. If you've ever shot lacquer over enamel and watched the surface wrinkle like a prune, you understand the need for a sealer.

Primer-Sealers: PPG's epoxy sealers are unique in that they serve a dual function, acting as both a primer and a sealer, in some instances they will also take one-step out of the finishing procedure. Again, primer/sealers are not designed to fill imperfections, so heavy, multiple coats are a waste of material. One light coat is all that's necessary.

Acrylic Enamel: At one time acrylic enamel was very popular, but it's seldom used today and will be phased out eventually due to tightening VOC regulations.

Single-Stage Deltron: Modern paint materials are available in different packages. There are single-stage urethanes that are applied very much like the once popular acrylic enamel—shoot on the color and you're done. PPG's Concept DCC is Deltron's direct gloss system, meaning no clearcoat is required.

Basecoat/Clearcoat: Another option is PPG's Deltron 2000 (DBC) system. With these two-stage paints a highly pigmented base is applied that looks flat when cured. A second stage, or clearcoat, provides the gloss as well as a durable, chip-resistant surface. When used with a basecoat system the clearcoat provides the gloss and as well as a durable, chip-resistant surface.

Envirobase: The Latest in topcoats. When solvent-based paint is applied a certain amount of VOCs are released. Interestingly, while the highest level of VOC release is during painting it continues while the paint cures. To comply with mandated VOC reductions waterborne paints were developed.

For some street rodders the mention of waterborne paint is a reminder of the first OEM attempts at using water-based materials and what a disaster it was; not only did it turn dull but it peeled off in sheets. However, today's waterborne paint uses totally different and vastly improved technology to reduce the release of VOCs while offering unmatched gloss and service life. Today the technology is used in about 70 percent of OEM colors. PPG's waterborne products use high-tech latex resin pigments in a special acrylic waterborne latex resin. Interestingly, one of the advantages to waterborne paint is the way the pigment stays in suspension—mixing machines and mechanical paint shakers are no longer necessary just shake the containers by hand and combine the required materials.

CHROME AND POLISHED ALUMINUM

As the body- and paintwork were being seen to at HRBD, Art Holman and Dave Dougherty at Sherm's Custom Plating were sorting out miles of stainless and aluminum trim that would be refurbished along with all the pieces that would go into the chrome tanks. From the factory the aluminum pieces were anodized, which provided an almost satin sheen. We were looking for a finish with a little more sparkle so the aluminum pieces were polished to the point of looking like chrome—they were then clearcoated by HRBD to preserve the finish. Like everything Sherm's does, the chrome, stainless, and aluminum components were finished to concours standards.

While all chrome shops use the same basic plating process, how parts are prepped has a huge impact on

3

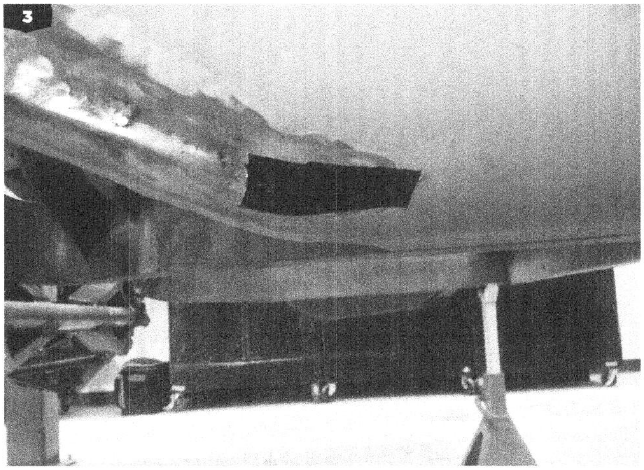

Rusted areas were cut far enough away to get into solid sheetmetal so a patch could be welded in place. Most of the damaged areas were small.

4

On the other hand some damage, like this hole in the front fender, were more extensive.

the finished product, just as it does with paint. In both cases the finished product is only as smooth as the surface below, and the smoother the surface the more reflective it will be. Like a paintjob, the prepwork for chrome is the time-consuming, and consequently the expensive part of the process.

Chrome is applied using a process called electroplating. During the process electrical current flowing from an anode carries dissolved metal in a solution where it is applied to the surface of cathode, or the part being plated. But while that sounds simple, in reality the plating process is a complicated requiring expensive equipment, constant monitoring of solution strengths and temperatures, not to mention strict adherence to a long list of safety and health standards. Sherm's is a zero-discharge facility, which means they have invested in, among other things, an ion exchange unit that continuously circulates and purifies process water, making it cleaner than what comes out of the tap.

When parts come into Sherm's for plating they are cataloged and inspected. Previously chromed parts are stripped and all parts are mechanically and chemically cleaned. Next the visible surfaces of the part are smoothed to remove any imperfections. If necessary some filling is done—lead is normally used for cosmetic fixes while structural repairs call for welding with the base material.

With the prepwork done the next step in plating is the copper coat, which is like primer for paint. In some cases several coats of copper will be applied and smoothed to make the surface perfect. On many early cars trim parts were made from a material often called pot metal. That

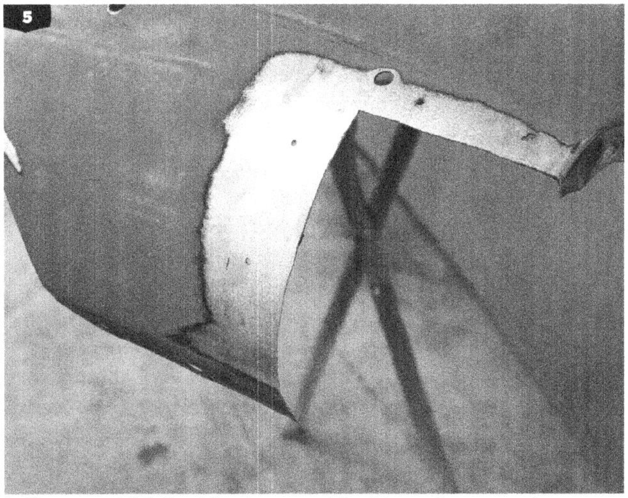

In the case of this front fender a considerable portion was cut away.

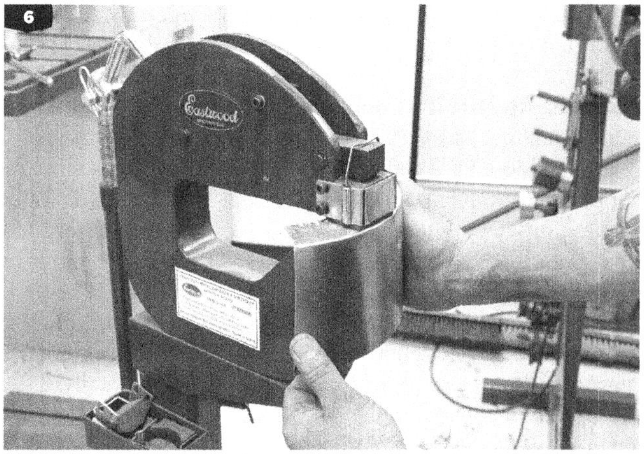

The HRBD crew used a variety of Eastwood tools, including this shrinker/stretcher used to form the patch panels. Here the front fender patch is being made.

This small panel has been formed to fill the hole above the wheel opening.

Here some of the patches have been welded in place. Note the plug on the small hole is just tacked. By using an Eastwood shrinker/stretcher a patch panel for the front fender was fabricated. After welding and grinding a skim coat of body filler would be applied.

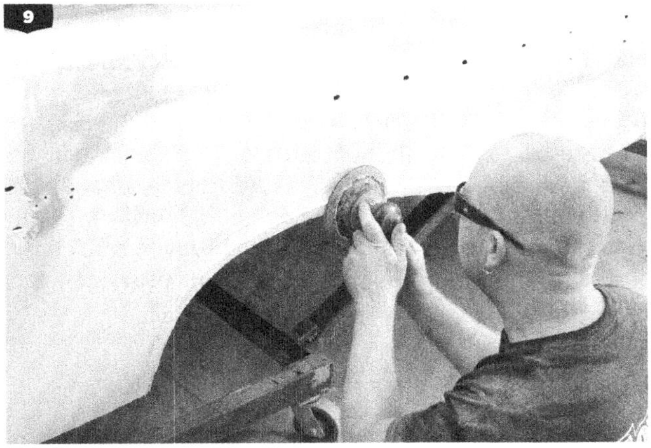

With the rear fender patch welded and ground a thin coat of body filler was applied after which the vertical surface was sanded flat.

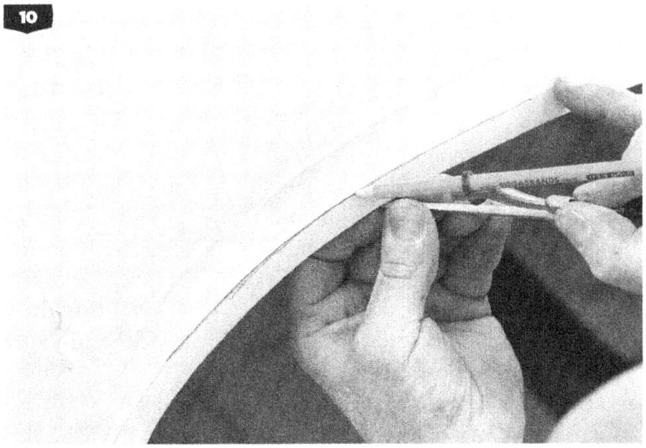

To shape the fender lip properly a compass was used to establish its width.

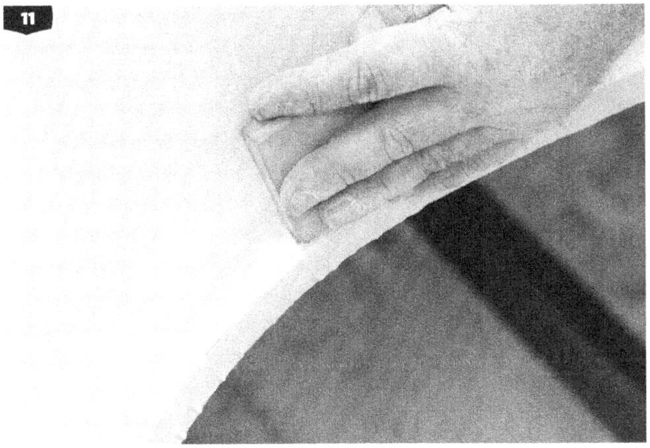

Once the proper width of the fender lip was established, masking tape was applied as a sanding guide. Careful sanding to the guideline established by a strip of masking tape resulted in a perfectly formed repair—neat trick.

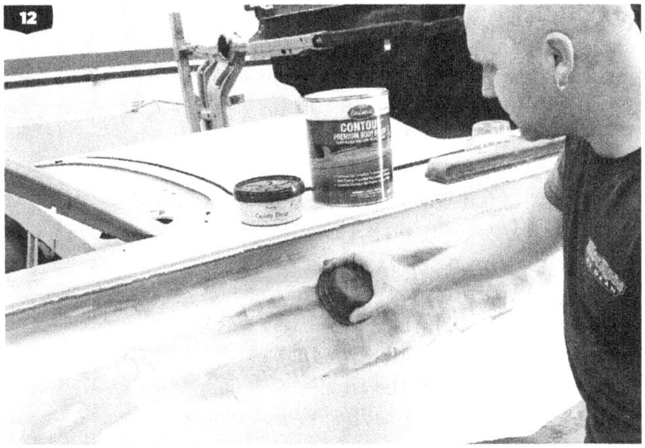

After the necessary rust repairs and hammer and dolly work were done the areas that were less than perfect were given a skim coat of Eastwood's Premium Body Filler. After an initial sanding Guide Dust was applied to indicate any highs and lows in the sheetmetal.

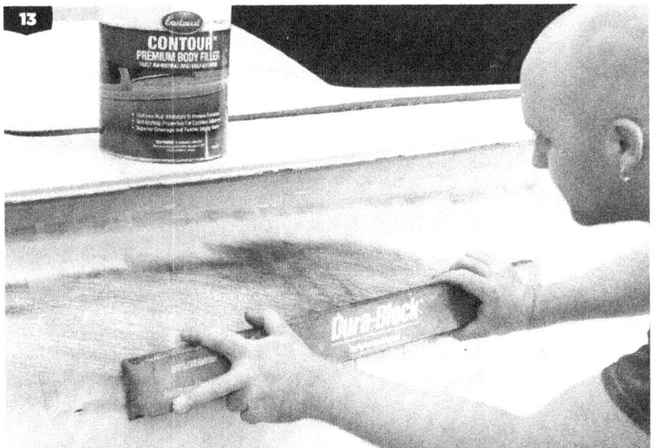

Here's HRBD's secret to primo bodywork—body filler, block sanding, primer surfacer, lots more block sanding, so on and so on. Note the long Eastwood block, perfect for the massive flanks of our Chevy.

We've found Eastwood's Rust Encapsulator to be a great way to combat corrosion and it results in an attractive matte black finish. It was sprayed inside the quarter-panels as a precaution. Every surface that could be reached was treated to a coat of Eastwood's Rust Encapsulator.

name came from the fact that "just about anything was thrown in the pot," including zinc, which has a low boiling point. During the casting and subsequent cooling process, air bubbles can be trapped in pot metal parts, which leads to corrosion and blisters under the chrome plating. The only way to repair a blistered pot metal part is to strip it, clean out the pit to get rid of the corrosion, then make the surface level using solder as a filler. The problem is lead won't adhere to pot metal, so the part is given a thin coat of brass. Although solder will stick to copper, there is a compelling reason for using brass. Intricate, ornate design work is often incorporated into the surface of these cast pieces and compared to copper a much thinner coat of brass can be applied to the piece so even the most delicate details can be preserved.

After the part to be chromed is smooth and straight nickel is applied. The equivalent of the first coat of color when painting, nickel provides a shiny reflective surface—but left unprotected nickel will tarnish. To protect the nickel the final step in the plating procedure is an application of chrome, its like the clearcoat over paint. Surprisingly, this is the least time-consuming step of all the processes, taking only a few minutes in most cases. But that doesn't mean there aren't some tricks involved. Depending on the shape of the piece being plated there may be high and low density areas—that means areas that are prominent are high density and will attract more current and consequently a thicker layer of chrome. Areas such as concaves are low density and won't receive as thick of a deposit. The cure for this is experience and specialized equipment.

Along with chrome-plated components stainless steel was used for trim pieces on '40 Fords and most other early cars. Over years it's not unusual for the trim to collect an assortment of dings and dents, but stainless is a tough and it can often be saved. Much like bodywork the procedure is to use hammers and dollies to remove dents, in some cases silver solder is used to repair rips and tears, all of which is followed by sanding and buffing. Properly done most stainless repairs are impossible to detect.

With body, paint, and trim work being attended to, a long list of components needed was sent to Classic Industries. They provided everything from the headlight buckets and hardware to the rear license plate trim and trunk latch and tons of stuff in between. Mirrors inside and out, an original-style antenna, assorted badges, trim clips, and more came from our friends at Classic Industries. *

15

All the paint products used on the AMSOIL/*STREET RODDER* road tour car came from PPG.

16

After a final application and blocking of K-38 primer the rich Ferrari Red color was applied followed by D894 high-solids clear.

A trip around the Chevy's sheetmetal with a buffer made the Impala sparkle. It looked spectacular in Dean's shop and is almost blinding in the sunlight.

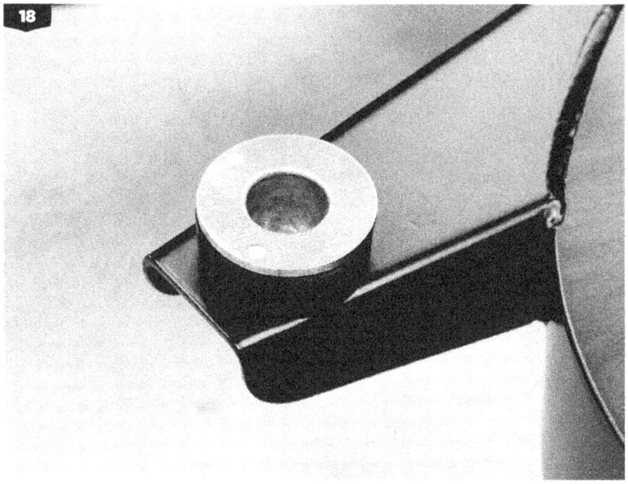

Before the body was dropped onto the Roadster Shop chassis new mounting pads from Classic Industries were put in place.

A collective sigh of relief could be heard when the painted body went from dangling in mid air to sitting securely on the chassis.

HRBD's attention to detail is astonishing—note the quality of the finish on the inner fender panel.

With the body secured, the HRBD team began hanging all the sheetmetal. ARP fasteners were used throughout.

Sherm's handled all the chrome plating (and there was a lot of it) as well as restoring all the aluminum trim (there was a lot of that too). Here Dave Dougherty strips the anodized coating off one of the taillight bezels.

After stripping Dougherty examined each part and began making repairs. Here a big dent in the taillight bezel has been identified.

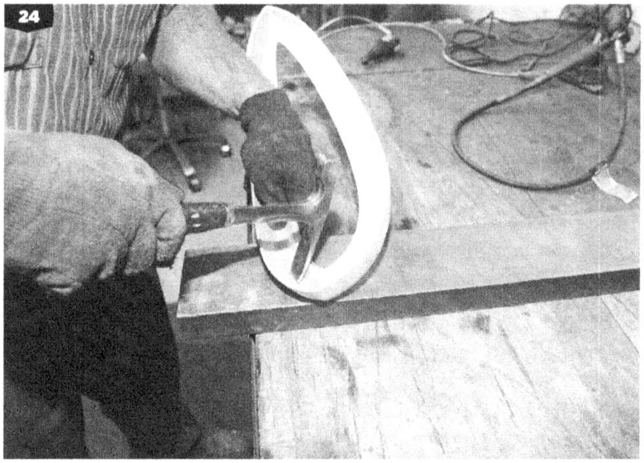

It takes talent, patience, and experience to remove the dents from aluminum trim—Dougherty has all that.

After removing the dents the parts are handed off to the polishers. This multi-step process involves a variety of sanding belts, polishing wheels, and compounds.

Sherm's spotless facility has the most up-to-date equipment available. That's what it takes to produce quality plating and meet all applicable environmental guidelines.

All parts previously plated or not are mechanically and chemically cleaned and stripped.

Copper is the primer in the plating process. Once the copper is applied the part is thoroughly rinsed in deionized water.

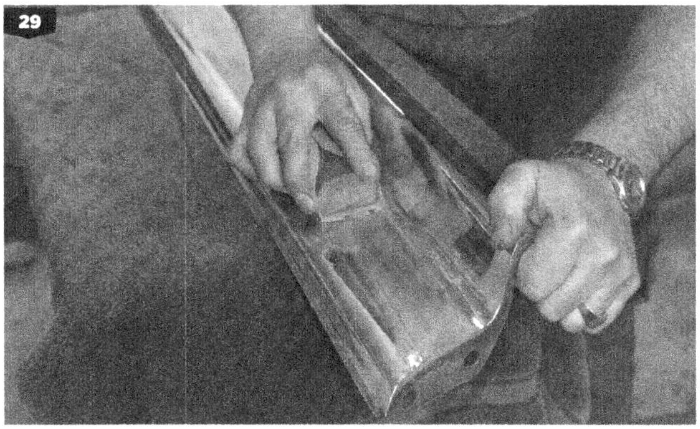

Art Holman demonstrates how to make chromed parts perfect. Just like paint making the surface flat makes it more reflective, copper is used as filler, and then the surface is block sanded until it all the highs and lows are eliminated.

Every part is closely inspected during each step of the plating process. Dave Dougherty takes a close look at a bumper fresh from copper plating.

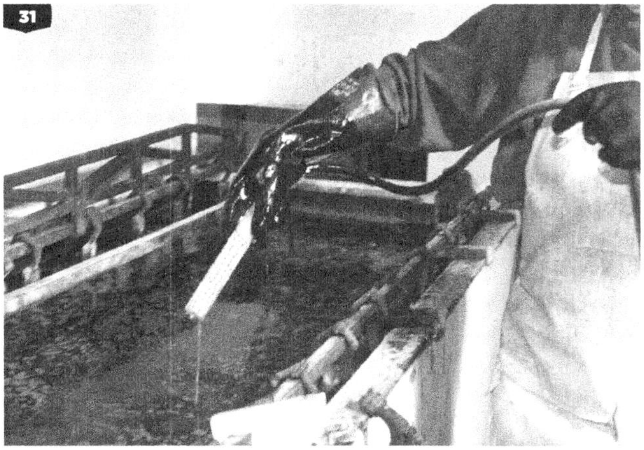

Once the copper-plated part is prepped it goes into the chrome tank. One of the tricks of the trade is to use a wand to direct current to low-density areas to ensure adequate material thickness.

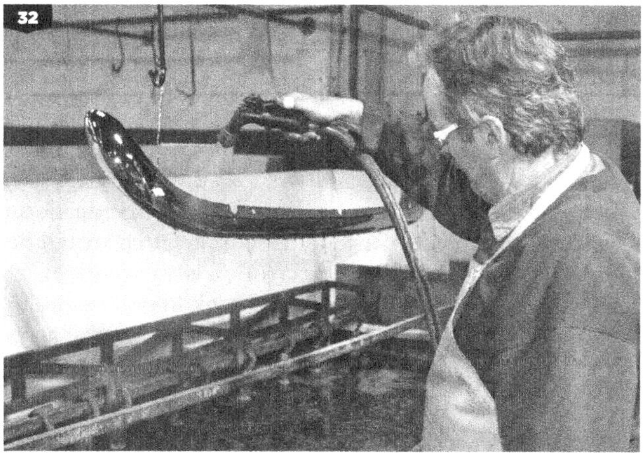

Out of the chrome tank the part is rinsed thoroughly and inspected. After one more cleaning the part is inspected again and then wrapped for delivery.

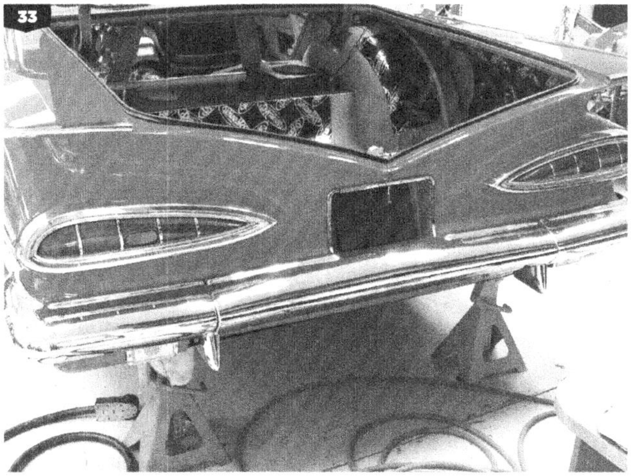

The results of Sherm's Custom Plating's efforts can be seen in the aluminum trim around the taillights, the chromed rear bumper, and the polished stainless trim on the fin.

Another of the trim parts from Classic Industries were the Impala and crossed flags' emblem found on the quarter-panels. Replacement door hardware from Classic Industries replaced the worn originals. With refurbished trim from Sherm's and the reproduction badges from Classic Industries, the Chevy sparkles.

CHAPTER TEN

In With the Good, Out With the Bad

Fuel and the Exhaust Systems

By Ron Ceridono | Photography by Brian Brennan

Due to the reconfigured Roadster Shop frame and rear suspension components used under our '59 Chevrolet the stock gas tank was deemed unusable and was tossed in the scrap pile. Of course even if the original tank had fit, using it would not have made any sense. Fuel tanks that have been empty for long periods are often rusted beyond use, and with today's ethanol-blended fuels corrosion is an even bigger worry. The smart move is to switch to a stainless steel tank like those built by Rock Valley for the Roadster Shop. Stainless won't corrode, the tank can accommodate an internal electric, and they are certainly better looking than the original. Along with the tank we used the Roadster Shop's stainless support straps.

One of the major advancements in automotive technology was the introduction of electronic fuel injection. By the end of

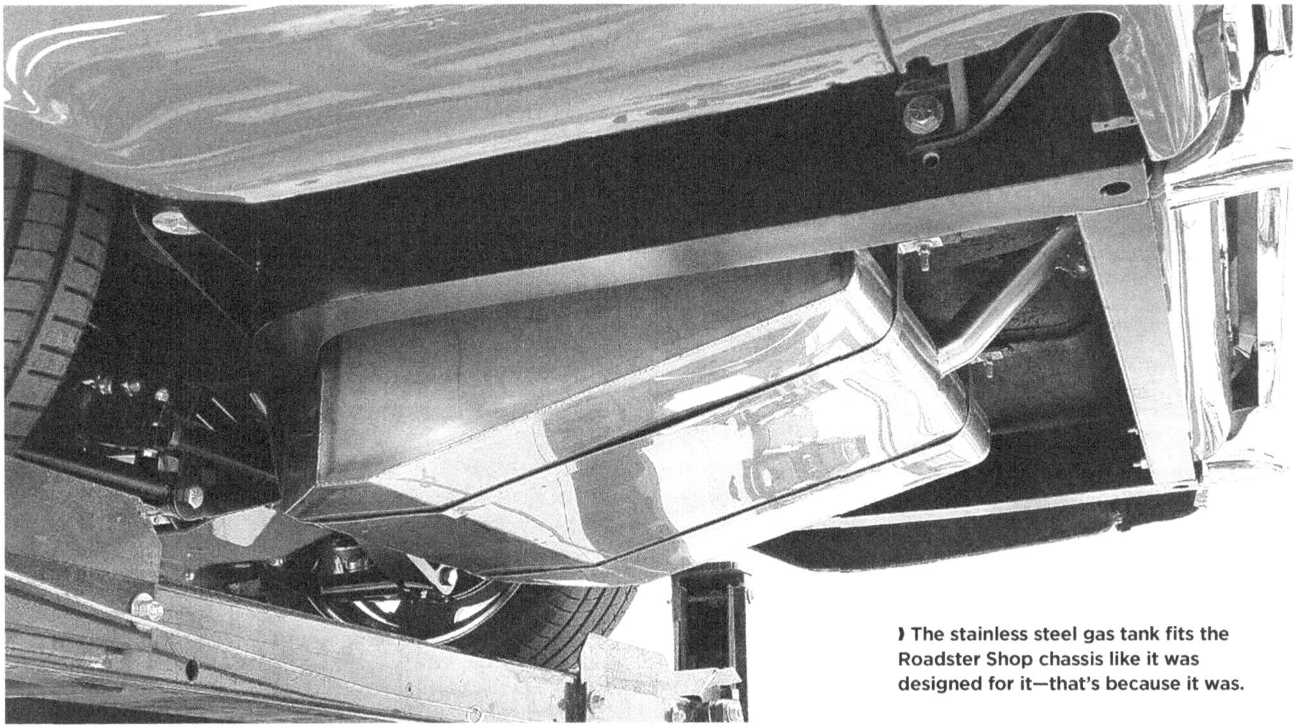

❯ The stainless steel gas tank fits the Roadster Shop chassis like it was designed for it—that's because it was.

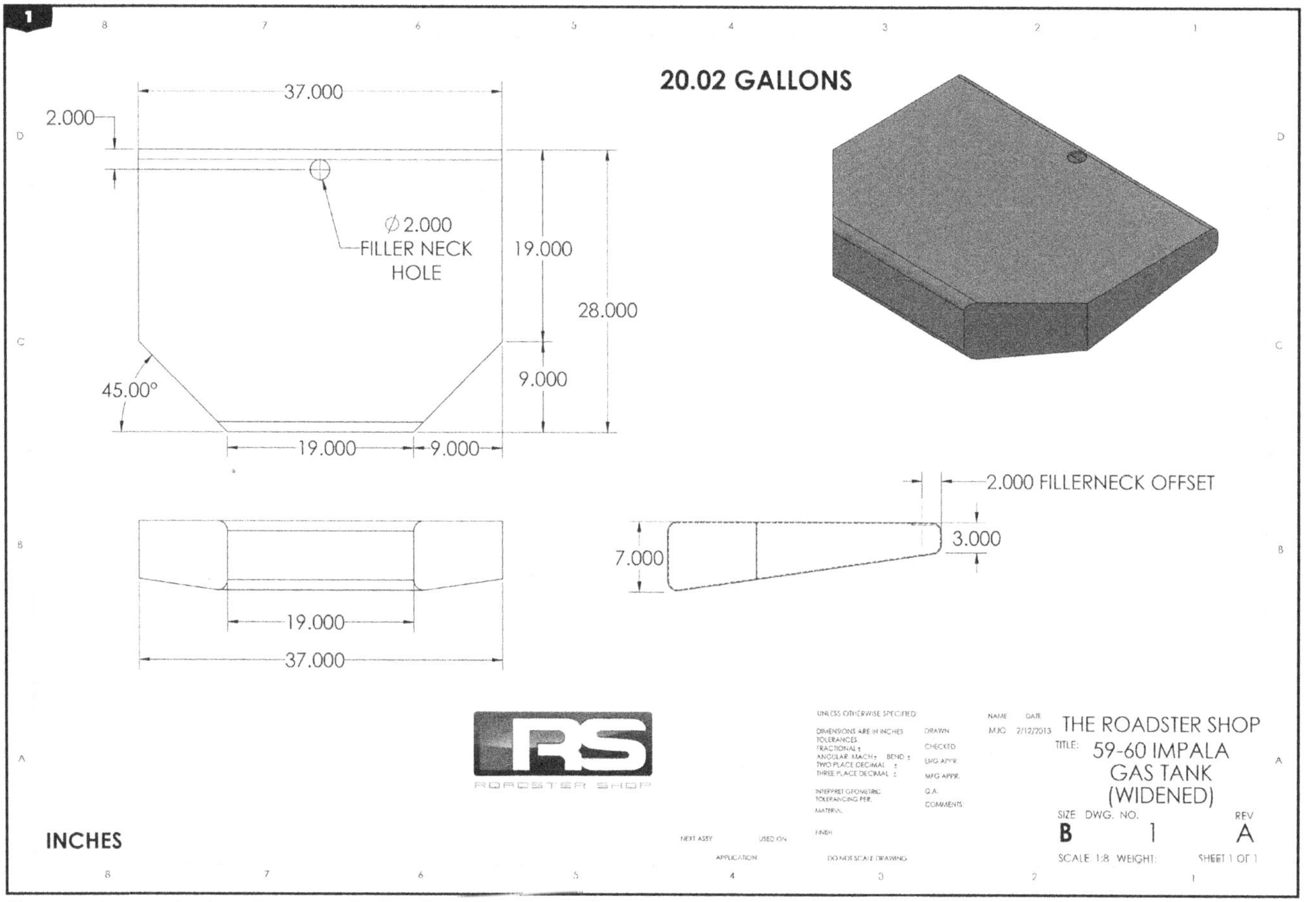

There are two gas tank options from the Roadster Shop, 17.7 gallons and 20.02 as shown in this drawing. Custom tanks for special applications are also available.

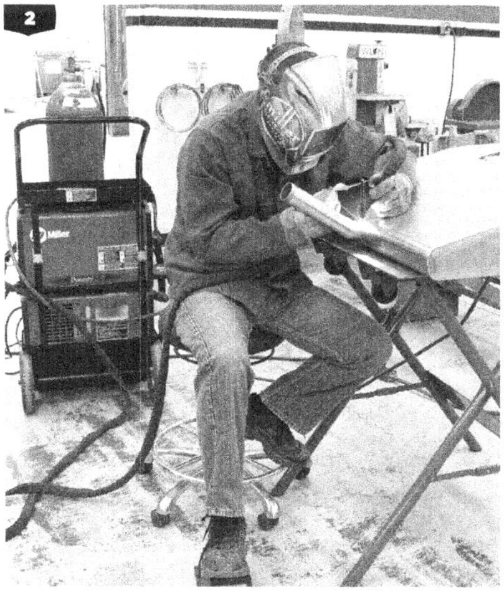

Here the filler neck is being TIG welded to the tank. As the filler is in the trunk a threaded sealed cap is used.

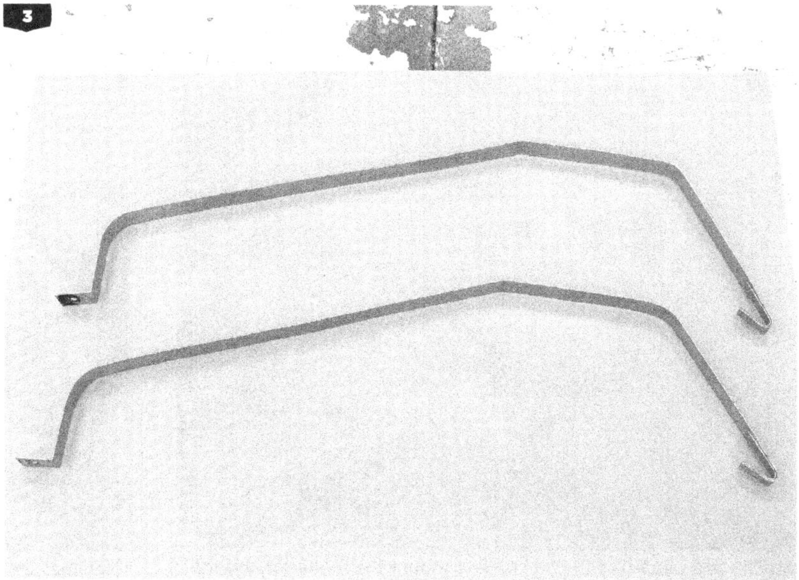

Stainless straps secure the tank to the trunk floor.

1990 carburetors had been just about eliminated from U.S.-built automobiles (there were a few minor exceptions, some of which were police cars). Of course it didn't take long for street rodders to adapt the new technology and with the readily available aftermarket systems making the change from a carburetor to fuel injection is easy to do.

While converting to fuel injection is simple enough, there is more to it than simply bolting components onto the engine and hooking up the wiring harness. Fuel injection requires considerably higher pressure than a carburetor so an appropriate electric pump is required. There are inline pumps for that purpose, but a much better alternative is Aeromotive's line of pumps that mount inside the tank. Mounting the pump in the tank has a host of advantages, chief among them is with the pump submerged in fuel it runs cooler, which translates into longer life and it's considerably quieter to boot. For the Road Tour Chevy, we used Aeromotive components, including a 340 in-tank pump pressure regulator, filter, braided line, and AN fittings.

Another in-tank option is Aeromotive's Phantom Stealth system. It comes with unique foam "basket" that serves as a baffle and a sump to ensure the fuel pump always has an adequate fuel supply. The kit comes with a unique drill jig and basket installation fixture that makes installing the basket assembly quick and easy.

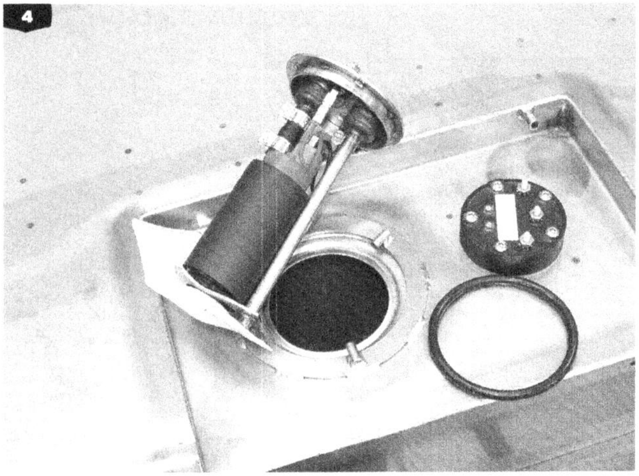

An excellent in-tank pump is the model 340 from Aeromotive, baffles in the tank are recommended to keep the pump submerged in fuel. The O-ring seals the pump assembly to the tank; the fuel level sender is to the righ

All the fuel line fittings are AN—on the left is the supply for the pump, on the right is the return from the regulator.

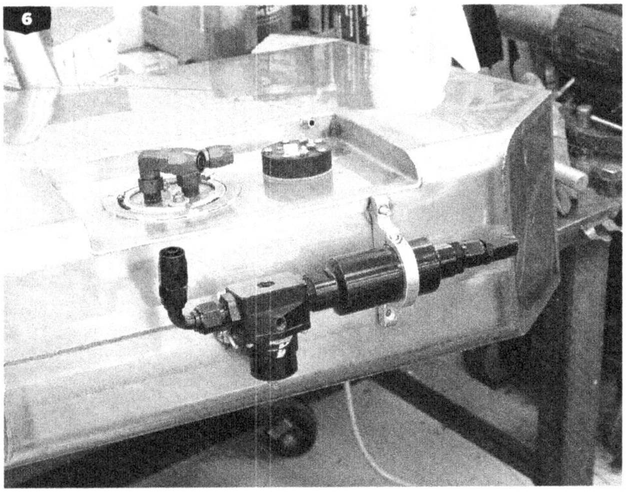

Studs were welded to the fuel tank to mount the regulator and the filter.

Fuel lines are braided stainless steel with lining material to resist ethanol-blended fuels.

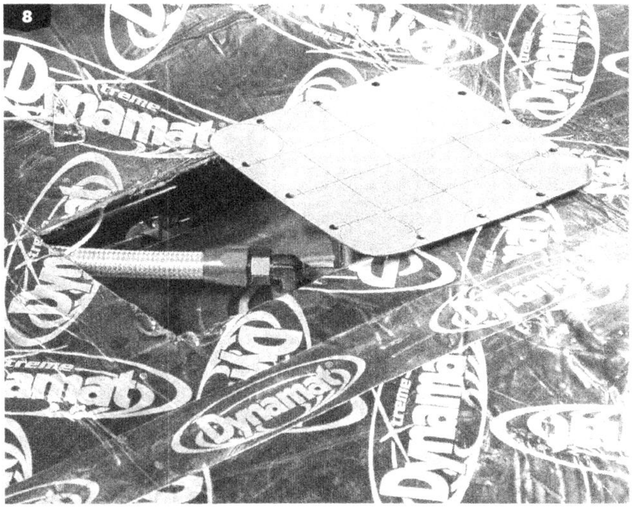

If access to the fuel pump or gauge sender is necessary Hot Rods By Dean included an access panel in the trunk floor.

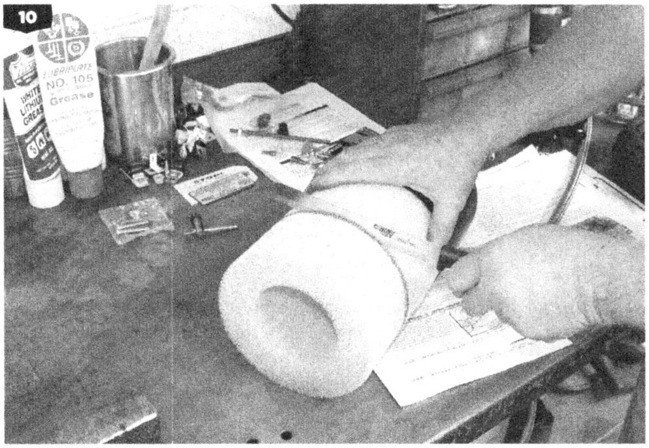

After measuring the depth of the tank, and adding an inch to provide some compression on installation, the foam is marked for trimming.

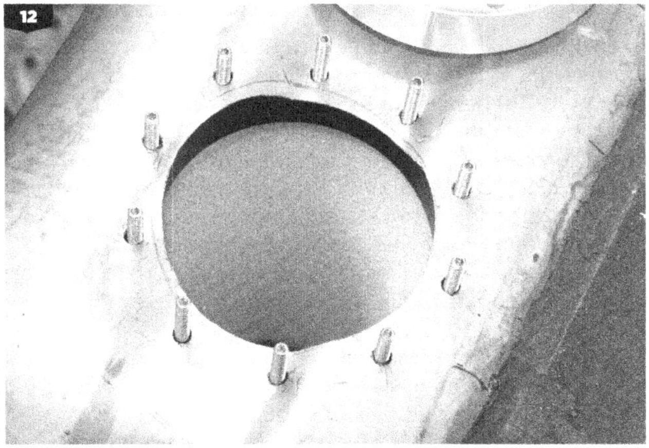

Once the basket is completely inside the foam springs back to the proper shape.

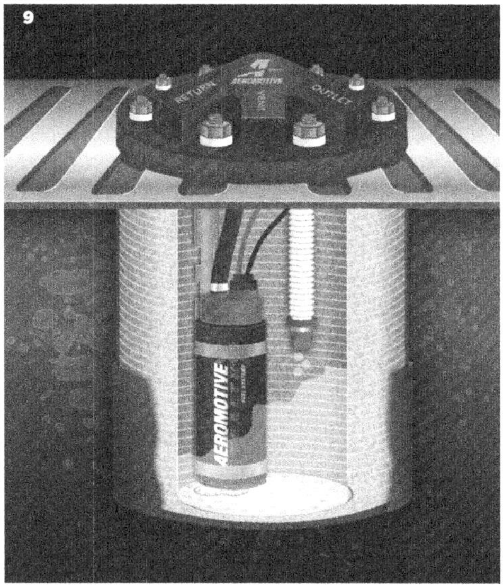

Aeromotive's Phantom fuel pump kit (PN 18688) includes this clever foam baffle and sump assembly. This illustration shows how the pump is positioned in the baffle/ sump assembly.

Starting with the sump portion, the basket assembly in pushed through the tapered portion of the drill guide.

13

A thick foam gasket seals the mounting assembly to the tanks and compensates any irregularities in the tank's surface.

14

Once in place the push-on connectors are attached to the fuel pump's feed and ground terminals.

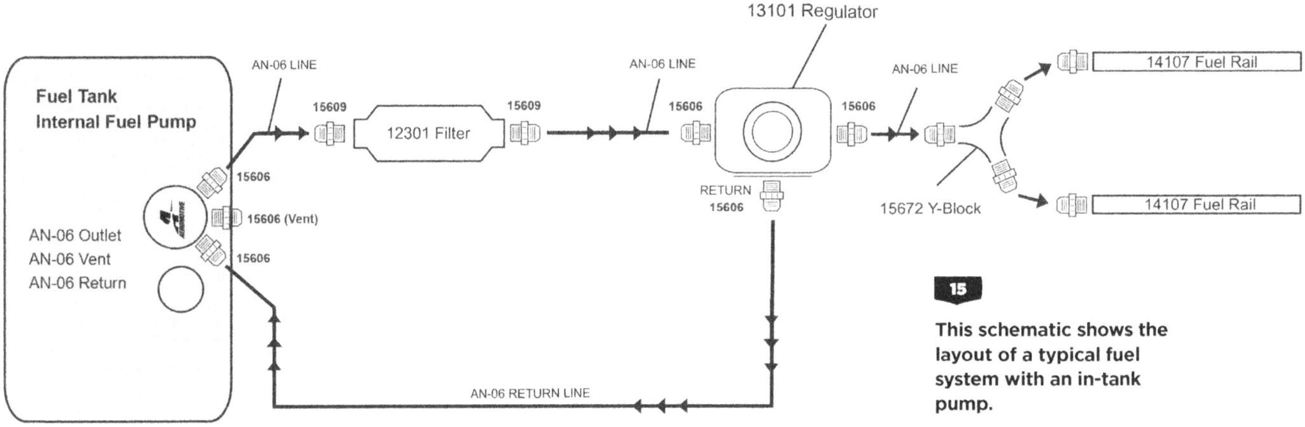

15

This schematic shows the layout of a typical fuel system with an in-tank pump.

Fuel Systems 101 from Aeromotive

Fuel systems fall into two basic categories: static or deadhead, and dynamic or return type. Static systems typically used with carburetors essentially deliver fuel "on-demand." Normally the fuel pump (mechanical or electric) will supply the carburetor's needle and seat with around 7 pounds of pressure and under normal conditions the pump will keep up with demand and maintain that pressure. As fuel is used by the engine the float drops and the needle valve opens, the resulting reduction in pressure causes the diaphragm in a mechanical pump to begin moving again, or the bypass in an electric pump to close, building the pressure back to 7 psi.

Dynamic, or a return-type fuel systems can be used on carbureted systems but are most often found on fuel-injected engines. With a return-style system the pump pushes fuel at 100 percent of its capacity all the time—it also pulls 100 percent of its capacity from the fuel tank all the time and that's why pumps in return style system are more sensitive on the inlet side (we'll discuss that further when we get to filters). With a return-style system fuel flows through the entire system at the pressure determined by a regulator—be it 7 psi for a carburetor or 60 psi for EFI. The regulator allows fuel to "bypass" and return to the tank, then as the engine begins to use more fuel, less returns to the tank.

EFI and modern engine technology have combined to create the ultimate street-driven race car. With ordinary combinations producing from 300 hp to as much as 500 hp, the stock EFI fuel system has proven flexible.

Fuel Filters

Filters are rated by the size of contaminates they will trap (a micron is 1/1000 of a millimeter or 0.00003937 of an inch). Aeromotive recommends that 100-micron stainless steel filter elements be used on the suction side of the fuel pump, but not all 100-micron filters are created equal. Another factor is the filtration area. Aeromotive's smaller filters have over 63 square inches of filtration media, ensuring there is ample room for flow and eliminating any pressure drop. It's not uncommon for a fuel pump to be ruined because the 100-micron fuel filter used only had 1 square inch of filtration media. The small filter area can create a huge restriction, which vaporizes the fuel causing cavitation that will lock up the pump. Common filters that cause these issues are those that use a disc-style screen. Always look at the micron rating and the filtration area. Both are equally important to the life of your fuel pump. When installing a fuel filter on the inlet of an Aeromotive fuel pump only use an Aeromotive 100-micron PN 12304, 12302, or an acceptable equivalent. On the outlet Aeromotive recommends PN 12301 or 12310 filter with 10-micron fabric for all Aeromotive pumps except the Pro Series EFI pump, which requires filter PN 12302 with larger stainless steel element and -12 inlet and outlet fittings.

Fuel Line FAQs from Aeromotive

Q: What size pump supply line should be used?
A: The size of the line from the tank to the pump's simple rule of thumb is: Do not make it smaller than the pump's inlet port size, and don't hesitate to make it bigger if you can. This is easy to figure out when the pump ports are ORB-AN type. For example, if the pump inlet port is ORB-10 use an AN-10 or larger fuel supply line, and make sure the tank or fuel cell outlet pick up tube (whatever the pump is drawing through inside the tank) is the same AN size. When you're dealing with NPT threaded ports, use the line size that is a minimum 1/8-inch larger than the NPT port thread, for example 3/8-inch NPT should have 1/4-inch (AN-08) lines, 1/2-inch NPT should have 5/8-inch (AN-10) lines. By the way, the subject of AN, NPT, and line sizes is fully addressed in Tech Bulletin number 901 for those who want more detail.

Q: What size feed line to the engine is appropriate?
A: From the fuel pump to the engine, in low-pressure (carbureted) systems, use a line equal to the fuel pump outlet port for optimum results. For fuel-injected engines, the above applies, but you may use a line that is one AN size smaller than the fuel pump outlet port and still get acceptable results.

Q: What about the return line and what are the rules there?
A: In order to determine the minimum return line size you must consider the actual system pressure needed (is it carbureted or EFI?) and factor in the chosen fuel pump's

Continued on next page

Hard-core engine combinations exceed this mark with the exotic making 1,000-plus horsepower on "the street." Applications above 500 hp universally require a complete fuel system makeover, from fuel rails to fuel pump. A key part of any fuel system upgrade is the fuel container itself. The debate is whether to modify the stock tank or install a racing fuel cell.

There are several benefits for retaining the stock fuel tank in a high-horsepower streetcar. It has a larger capacity than most fuel cells, already has a mounting location and hardware, has provisions for filling from outside the vehicle, has a cap that both vents and seals, and is already on/in the vehicle.

The drawbacks to stock fuel tank retention are more numerous but less obvious. The stock pick-up/pump assembly is restrictive, requiring complete replacement with a fabricated assembly. When using a stock tank with fabricated pickup, unless the fuel level in the tank is 3/4 full or higher, the internal well, which the stock pump draws from, is far too small and poorly supplied with fuel from the rest of the tank. Faced with the demand of a large pump, drawing through a fabricated pickup, it has no chance of refilling fast enough to support WOT full engine load. Under low demand, cruise-type conditions, the large volume of fuel delivered to the rails is unused and returned. The same fuel, picking up heat from the pump and the rails, is constantly recycled to and from this well, rapidly increasing fuel temperature. Common problems associated with stock fuel tanks and fabricated pickups are pump cavitation, vapor lock, varying fuel pressure, exaggerated pump wear, and lean conditions during both low and high loads.

Fuel Line FAQs from Aeromotive

output-flow. If we could, let's put one myth to bed here, there is NO "rule of thumb" relationship between the line size feeding the regulator from the pump and the line size that should be used to return from the regulator back to the tank. There are a couple of factors and one rule that can help here. Factor One: the lower the system pressure will be, the larger the return line must be (so if you're converting from EFI to carbureted, be prepared to install a larger return line, sorry!). Factor Two: the higher the fuel pump flow rate, the larger the return line must be. And The Rule: The return line size must be large enough to handle the flow from the pump while creating at least 3 psi less back pressure than the regulator is set for.

As you may have noticed, none of this has anything at all to do with the size of the supply line. The single biggest return line mistake we see, and it comes up every day, is the person who has reduced the return port in the regulator to a smaller return line. Having done this, the call comes in to report a defective regulator because they can't get the pressure to go down. What is the foolproof rule? Treat the regulator return port exactly as you do the fuel pump inlet port ... You can always make the return line bigger than the regulator return port, just like you can make the suction line bigger, and contrary to what you may have heard this will have NO negative impact on pressure control. Of course, if you do make it smaller you run the risk of creating too much return restriction, resulting in a regulator that won't adjust and causing false-high fuel pressure, carburetor flooding, driveability problems, etc. Note: When using the regulator return port size to determine the size return line to run, the same NPT to AN rules mentioned above apply.

Q: What are some general rules to follow when routing fuel lines? It's common sense to keep them away from moving parts, but what other not-so-obvious mistakes should be avoided?

A: This one's pretty easy. It's best to keep the lines secure. That can be one of the most common pitfalls made. Properly secure the lines to ensure that they do not move and incidentally rub or bump a moving part. The other side of coin would be to try and avoid heat sources wherever possible. Keeping the fuel cool is important to the system and it will pick up enough heat as is with the radiant heat from the pavement and the engine bay. Avoid placing them next to exhaust pipes, etc.

Exhaust

IN 1983, California Sprint Car racers were experiencing "excessive sound" issues at numerous racetracks and were told to quiet the cars down or lose racing privileges. All attempts to silence the race engines without sacrificing power repeatedly failed. Ray Flugger applied his 45-plus years of exhaust design experience, and out of an 800-square-foot barn in Kenwood, California, came Flowmaster's first patented design of racing silencers. Flugger attached handles to ease carrying the square-shaped race mufflers through the race pits, and the mufflers became known as "suitcase mufflers." These innovative mufflers were the result of years of development with top race engine builders, thousands of hours of dyno research and "real-world" on-track testing. It wasn't long before Flowmasters found their way to the street where that very identifiable rumble was often heard.

In response to requests for components from shops and do-it-yourselfers, Flowmaster has developed a number of kits to make building a custom exhaust system easier than ever.

Flowmaster Kits

Outlaw Series: feature race-bred components with only slight reduction on sound over open exhaust tubing on non-catalytic converter-equipped vehicles. With new vehicles featuring catalytic converters, Outlaw Systems can be compatible with regular street driving. This system provides the ultimate in aggressive-sounding performance.

American Thunder Series: has the classic sound of a Flowmaster chambered exhaust, which are tuned to deliver more power, torque, throttle response, and mileage. These exhaust systems deliver that authentic Flowmaster sound.

American Thunder Pro Series: deliver that same great aggressive exhaust note as the American Thunder chamber exhaust system, but with a deeper, uniquely special tone due to the use of our laminar flow muffler technology.

DBX series: features Flowmaster's newest laminar flow technology muffler. The DBX delivers the same attributes as the American Thunder Pro Series mufflers with a deep tone and great overall acceleration sound, but in a compact case perfect for smaller vehicles. The DBX features 304 stainless steel casings so they can be polished.

Force ll Exhaust System: perfect for those looking for increased performance and tone enhancement, but are concerned about installing an exhaust system that is too

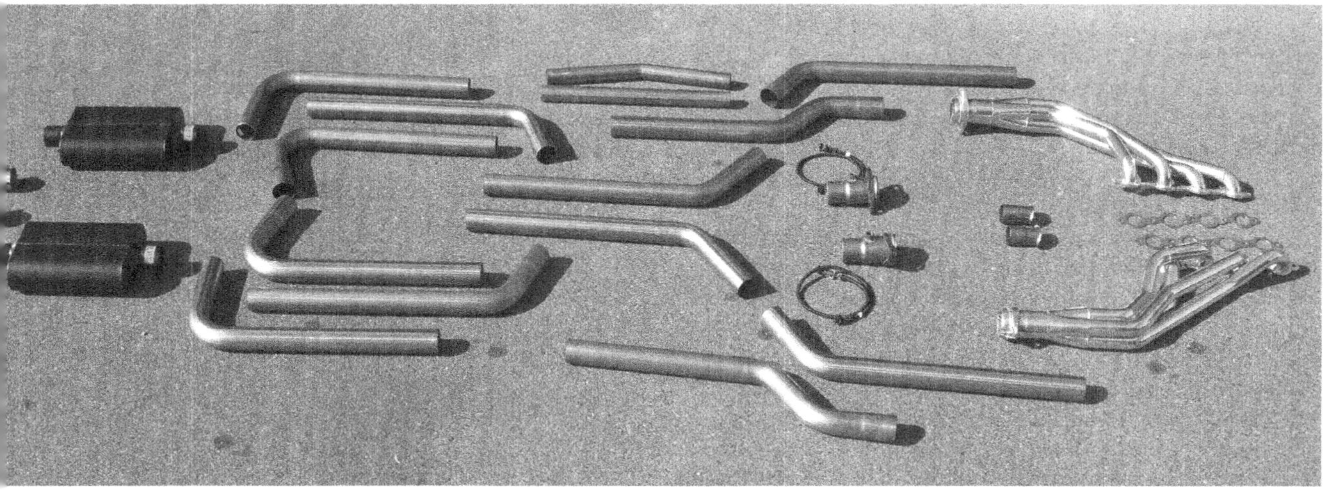

Hot Rods by Dean used headers, mufflers, pipe kit, and stainless tips from Flowmaster to fabricate the exhaust system.

Flowmaster's Scavenger Series Elite long-tube exhaust headers are designed to replace the restrictive factory cast-iron manifolds. These high-quality exhaust headers are manufactured from stainless steel and are ceramic coated for maximum durability. They feature 1-5/8 inch primary tubes, 3/8-inch thick flanges, and incorporate Flowmaster's popular 3-inch ball flange collector connections ensuring a leak-free installation. Scavenger Series long-tube exhaust headers deliver maximum performance and are suited to the enthusiast who wants the best.

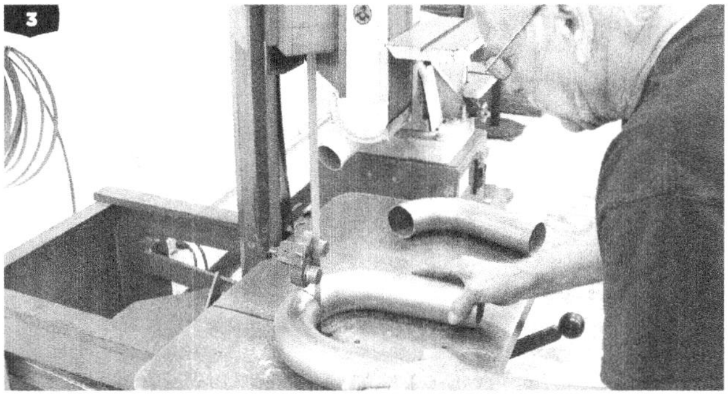

The Flowmaster kit incudes mandrel bends that can be cut as needed.

There's nothing quite like the sound from Flowmaster mufflers. In this case the mufflers were positioned, then the pipes were created to fit.

Due to the low stance and the independent rear suspension it took some planning to get the pipes to the rear of the car.

aggressive for daily street driving. The Force II exhaust systems commonly incorporate larger displacement mufflers and resonators to control overall sound, but increase the "performance impression" made by the vehicle.

Mandrel Bends

All Flowmaster kits are manufactured with mandrel-bent tubing for smooth high-flow bends, which are not only aesthetically pleasing but speeds the exit of waste exhaust gases unlike compression-bent tubing found from other manufacturers or muffler shops. A typical compression-bent tubing exhaust system can easily limit exhaust gas flow by 40 percent, which not only robs your engine of torque and horsepower but also wastes fuel and creates unwanted tuning resonances in the exhaust system.

The pipes leading from the mufflers snake their way under the rear crossmember. Note the hanger attached to the chassis.

A bung for the FAST EFI computer's oxygen sensor was welded into the coupler that connects to the header collector on the right side of the engine.

Once past the rear suspension the pipes were routed outside the framerails.

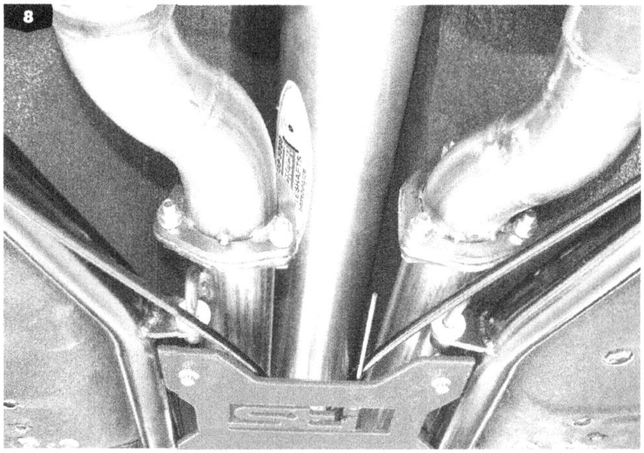

To keep them from dragging, the exhaust pipes were tucked into the tunnel with the driveshaft. Note the flanges that allow the system to be easily installed/removed.

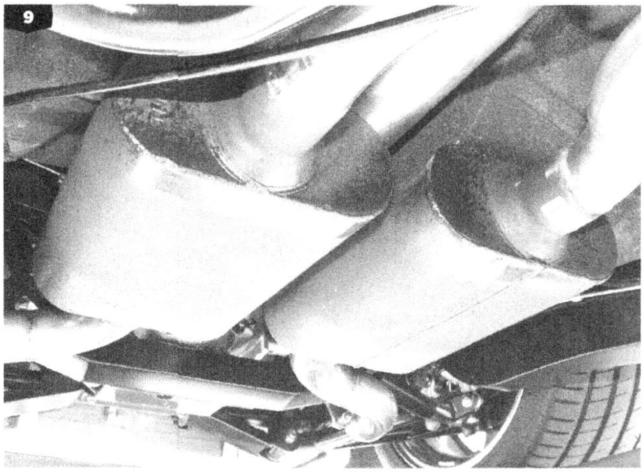

As cramped as the quarters are under the car, the mufflers are still above the bottoms of the framerails so they are hidden from view—but their rumble lets everyone know they are there.

Balance Pipes, H-Pipes, X-Pipes

An H-pipe, commonly referred to as an equalizer, balance, or crossover pipe, connects the two exhaust pipes. It will balance the pressure of the exhaust pulses in the system, thereby reducing interior and exterior sound while increasing power. In general, the H-pipe should be installed as close as possible to the header collector. With manifolds, the H-pipe is placed just past where the downpipes become horizontal. H-pipes should be the same diameter as the main pipes or no more than 1/2-inch smaller. Flowmaster also offers a Universal Scavenger crossover pipe that is great application in any high-horsepower racing applications. It uses D-Port Tuning technology, which helps with scavenging of waste exhaust gas and builds torque.

Tailpipes

Tailpipes aid in keeping atmospheric pressure from rushing back into the muffler. Tailpipes can improve performance and control sound. In addition, tailpipe length can affect the power as well as sound. Tailpipes should be a minimum of 10 inches, and not longer than 6 feet. Tailpipe length of more than 6 feet may need a resonator.

Positioning of Mufflers

The preferred location muffler is midway in the system, 18 inches back from the header collector. Other reasons to put them there is for improved interior resonance and sound control.

Heat Control and Shielding

Heat shielding may be needed where exhaust pipes pass close by electrical, fuel, or brake lines as temperatures in the exhaust system can exceed 1,200 degrees F. For heat control, Flowmaster offers both tailpipe and muffler head shields.

Suggested Pipe Sizes

The size of pipe used in an exhaust system is a critical item to consider. Pipe diameter will affect the sound level and performance characteristics of the muffler, but keep in mind that bigger is not always better. Too large of a pipe can actually hinder exhaust scavenging by allowing atmospheric pressure to move back up the pipe. As a rule, most street applications use 2.00- to 2.50-inch pipe, while modified street applications will generally use 2.50 or 3.00-inch.

10

To finish off the exhaust system HRBD added a set of stainless tips.

Jet-Hot Coatings

Jet-Hot explains the technology and science behind header and engine coatings, and the benefits they offer in power and durability.

By Stephen Kim | Photography by the Author

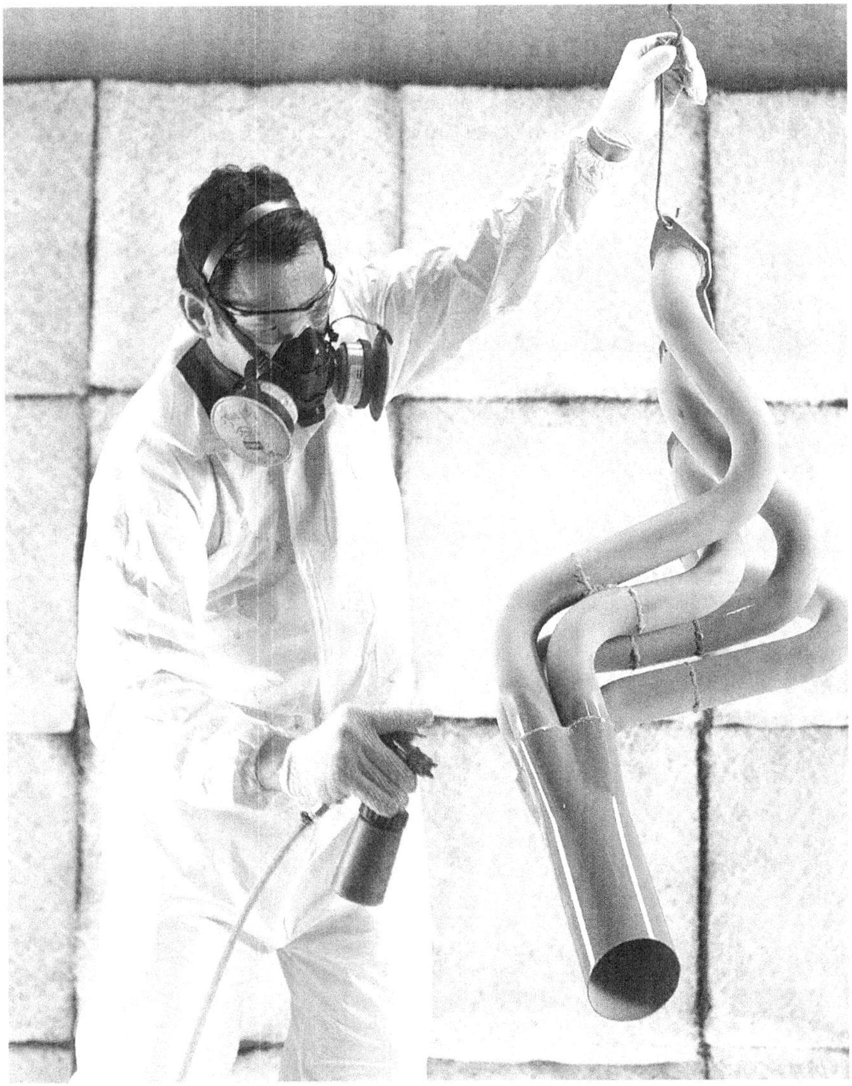

IN the walk of the hot rodding aftermarket, the term battle tested serves merely as a euphemism for track-tested racing hardware. When the products at hand are Jet-Hot's well-known ceramic coatings, however, that euphemism transforms into a literary truth. Before entering the automotive market, Jet-Hot earned its red, white, and blue stripes developing coatings for the U.S. military. Whether the challenge at hand was providing a thermal barrier on fighter jet engines and submarine parts, or protecting the launch gear on aircraft carriers, Uncle Sam turned to Jet-Hot to get the job done during the heat of combat. Recognizing the benefits of Jet-Hot coatings, the Army began using them on armored vehicles as well.

Eventually, Jet-Hot pondered the benefits of applying its proven coating technology to automotive applications. After extensive research, the company developed coatings specifically for automotive use, and racers and enthusiasts alike quickly realized the benefits of Jet-Hot's ceramic coatings. In the past, header coatings were nothing more than aesthetic products that provided little to no thermal

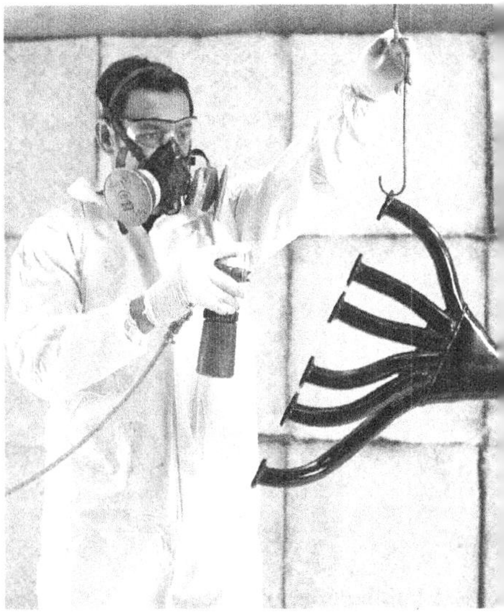

protection, but Jet-Hot changed all that by offering coatings that not only prevented corrosion, but also looked great. In recent years, engine builders have realized the benefits of applying Jet-Hot's thermal coatings on internal components such as pistons and cylinder head combustion chambers. Furthermore, the company has also developed a complete line of antifriction and oil-shedding coatings. To get the full scoop on coating technology, and how hot rodders can use them in their own rides, we sat down and chatted with Jet-Hot's Dave Burton. Here's what he had to say:

Origin of Coatings

Coating technology was first applied to headers in 1981. The technology was an adaptation of similar coatings Jet-Hot developed for the aerospace industry as a means of protecting high-temperature metals from oxidation. The extreme heat created by fighter jet engines, the corrosive environments they operate in, and the countless heat cycles they're subjected to caused very rapid parts wear, so the military turned to Jet-Hot to develop coatings that extended component longevity. Furthermore, Jet-Hot has supplied proprietary coatings to protect the parts on aircraft carrier launch systems for the U.S. Navy, as well as coatings for the U.S. Army's armored vehicles. Leveraging upon years of experience learned from providing long-term protection from corrosion and thermal fatigue in military applications, Jet-Hot transitioned into the automotive aftermarket in the early '90s. These coatings are similar, but have been updated and enhanced over the years to meet the specific needs of the automotive segment. Jet-Hot is constantly developing and testing new coatings for specialized applications and solutions. Customers can rest assured that Jet-Hot uses the same technology in our automotive applications as in our military applications to provide a superior level of protection.

Automotive Transmission

Given Jet-Hot's extensive track record in heavy-duty military applications, we began exploring how to apply our coating technology to address the heat management issues in hot rods and race cars. After releasing our first automotive products, Jet-Hot became the number-one coating brand on the market within five years. Enthusiasts also appreciated the high-sheen finish provided by our coatings. In the early '90s, Jet-Hot became the only coating company to support racers at the track. Legendary NHRA Funny Car champion John Force recognized the safety benefits of Jet-Hot coatings early on, as they substantially reduced header temperatures. His crew chiefs discovered that Jet-Hot's coatings allowed for quicker cool down periods, allowing them to rebuild their race motors more quickly between rounds.

Coating Makeup

Ceramic is a rather generic term that doesn't necessarily describe the makeup of a coating. All coatings are not created equal, and several elements go into Jet-Hot's coating that give it its heat and corrosion resistance properties and distinguishes it from the competition. At a glance people only see the aluminum part of a coating. Ceramics in general are compounds that are inorganic and nonmetallic. Since organic compounds contain carbon that burns under extreme heat, eliminating them from a

coating is very important. This is what makes ceramic coatings different from a powdercoat or epoxy, both of which have carbon bonds that start breaking down at 600-800 degrees F. Additionally, ceramics contain binders that hold everything together. A coating's filler material can be a pigment, paint, or aluminum. Using aluminum in our coatings allows polishing it to a high-sheen finish that enthusiasts are familiar with. The end result is a coating that not only offers high-temperature and corrosion resistance, but also looks great.

Coating Process

Jet-Hot's ceramic coatings can be applied to any set of headers, both old and new, and are backed by a lifetime warranty. After receiving the parts at our facility, they are inventoried into our computer and scribed with the order number on the pipes themselves. We also take a photo of every incoming order for referencing purposes. The next step is cleaning the parts. Usually a thermal degrease or chemical bath works well for light oils, oxides, and other contaminants. Internal engine parts, like pistons, are hand-cleaned with solvents or by using an ultrasonic process. The parts are then grit-blasted to remove scale and debris, and to get the parts down to a clean "white" metal surface. This gives the coating more surface area to grab onto, and ensures good bonding. Once the metal has been cleaned, the ceramic coating is applied using a spray gun similar to those used by body shops. After curing in an oven, the parts may get coated again, if necessary, before they are submerged in a tub of polishing media. The polishing stones vibrate across the metal surface to give the parts a nice sheen, and the parts are then hand-wiped before moving on to final inspection. From start to finish, the coating process takes about two days.

Inside and Out

It's not uncommon to only coat the outside of a header, but at Jet-Hot we feel it's important to coat the inside of the header tubes as well. There are several advantages of doing so, which is why we coat all headers on both the inside and outside. Since water vapor is a combustion by-product, as a motor and exhaust system cools down, water condenses and rusts a header from the inside out. To prevent this, Jet-Hot has developed special fixtures and techniques to apply our coating to the inside surface of header tubing. Regardless of how complex the inside of the tubing may be, we have developed methods to apply coatings to them. Additionally, this yields the benefit of reflecting heat away from the inside surface of the pipes. By reducing the amount of heat transferred through a pipe from the inside out, it keeps the outside header surface cooler as well. Superior heat retention also assists in improving airflow and exhaust scavenging. Furthermore, by ensuring that the heat travels more uniformly across the header surface, it eliminates the potential for heat to build up in localized hot spots. Jet-Hot does not charge extra for this work and provides a full interior

coating on all parts. Coupled with a full lifetime warranty, Jet-Hot stands behind its coating and workmanship.

Types of Coatings

Jet-Hot offers several different coating options: Extreme 1300, Extreme 2000, and Extreme 2500. Extreme 1300 is by far our most popular coating. It is easily identified by its polished aluminum finish, and is capable of providing continuous protection to a skin temperature of about 1,300 degrees F. This makes it ideal for naturally aspirated motors, and it can be applied to the majority of parts that we see. The next step up is our Extreme 2000 coating; it contains stainless steel elements that provide thermal protection to skin temperatures of about 2,000 degrees F. This type of coating is best suited for applications with turbos or superchargers that run very hot. Finally, Extreme 2500 is a full ceramic material that provides the best level of insulation for any application. As its name suggests, it can withstand skin temperatures of about 2,500 degrees F. It contains no metal content at all, can only be applied to the outside surfaces of parts, and is often used on components located downstream of turbochargers.

Turbo Applications

Many people are already familiar with the benefits of ceramic header coatings in naturally aspirated engines, but they offer advantages in turbo applications as well. Due to the extreme heat generated by turbo motors, Jet-Hot recommends using our Extreme 2000 or Extreme 2500 coatings in turbo applications. Coating the hot side parts can maintain the heat inside components better for improved thermal efficiency. Furthermore, our coatings help spread the heat out more uniformly. Interestingly, in rear-mount turbo combinations customers have reported reduced turbo lag when using Jet-Hot coatings.

Coating Maintanance

One of the biggest benefits of a coated header is its shiny good looks, but as the miles and years tack on, regular maintenance is required to prevent dulling and staining. Jet-Hot recommends that customers periodically clean their headers to remove any salt or road grime that can accumulate on the surface. The best way to do this is with light soap and cool water. Oil, transmission fluid, and engine coolant can be easily removed with fine steel wool or a Scotch-Brite pad. Any nonabrasive polish can then be used with a soft cloth to bring out a high-luster finish.

Aluminum will cloud up when cleaned in a hot condition, so it's imperative to allow the surface to cool down first. If headers are dulling, the coating is getting too hot for the application, or there is residue that's causing the color change. Furthermore, uneven fuel distribution and improper spark timing can result in extreme exhaust gas temperature that may exceed the temperature range of a coating. To avoid this, it's important to make sure an engine is in optimal running condition. Jet-Hot customers can also rest assured that their header

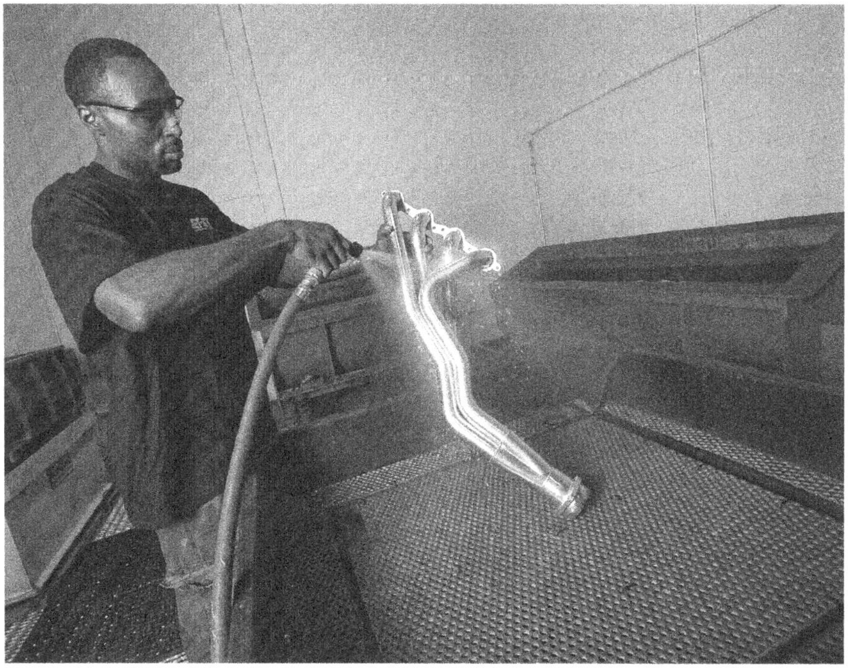

coatings are covered by a lifetime warranty. If there's ever a problem, just ship the headers in and we'll re-coat them under warranty.

Internal Thermal Coatings

With racing classes as competitive as they are these days, engine builders are increasingly using coatings on internal engine components to search for extra horsepower. In addition to exhaust coatings, Jet-Hot also offers coatings for internal engine components such as pistons, crankshafts, and bearings. Internal engine coatings fall into three categories: ceramic metallic thermal coatings, friction reduction coatings, and oil-shedding coatings. Ceramic coatings are often applied to piston crowns, valve faces, and the cylinder head combustion chambers. This provides thermal management and helps retain heat in the combustion chamber for improved power.

Antifriction Coatings

Friction reduction coatings provide improved lubrication and cooling to extend the life of internal engine components. On bearings and piston skirts, fluoropolymer coatings are designed to help maintain the oil film on the surface. The benefit is reduced friction, improved longevity, and increased horsepower. The coatings are very effective, yet thin enough that they don't impact bearing or piston-to-wall clearances at all. While the purpose of ceramic coatings on the valve faces is to maintain heat in the combustion chamber, an antifriction coating can also be applied to the valve stem. The resulting friction reduction can help prevent engine damage caused by sticking a valve.

Oil-Shedding Coatings

Engine oil not only lubricates parts, but it also carries heat away from internal components. Oil-shedding coatings allow oil to perform its job by carrying that heat away more efficiently. As such, oil-shedding coatings are typically applied to valvesprings, and the inside surfaces of valve covers and oil pans. The more oil these components can shed, the more heat that can be carried away. Other favorable characteristics of these coatings are that they're highly resistant to chemical acids and bases, and improve oil drain-back and oil management.

Cooler Air Temps

Keeping the temperature of the intake air charge as cool as possible results in a denser air/fuel charge, which increases horsepower. A cooler air intake charge is also more resistant to detonation. While it's true that header coatings reduce underhood temperature, and therefore the heat that soaks into neighboring components, the intake manifold itself can also be coated. Applying a ceramic coating on the bottom surface of the intake manifold mitigates the amount of heat that will transfer from the hot engine and engine oil into the intake runners. This keeps both the air and fuel cooler. The inside of the intake runners can also be coated, which sometimes improves the laminar flow characteristics of the induction system. In the near future, Jet-Hot plans on conducting extensive dyno testing in-house to measure the performance benefits of our engine coatings.

Coatings and Weldings

Since there are often clearance issues with headers, they sometimes need to be cut, welded, or modified for proper fitment. For the sake of simplicity, it is always best to complete these modifications or repairs before sending a set of headers in to get coated. This eliminates the possibility of having a section of tubing where the coating is no longer present. In the event that a section of tubing

that has already been coated needs to be welded, the coating must be removed with a grinder. Likewise, the heat of the weld must be increased as well.

Coatings and Dynos

When driving down the road, there is always a steady stream of air that removes heat from the engine compartment. However, this isn't the case on an engine or chassis dyno. The stagnant air in a dyno cell combined with the heat generated by sustained high-rpm engine operation can exceed the limitations of the aluminum in the coating matrix. This can ultimately cause dulling of the coating surface. To avoid this potential problem, it's best to use a large fan to help dissipate the heat of the engine and headers during dyno testing.

Jet Hot offers a variety of coatings:

Jet-Hot Extreme 1300 series is the most popular coating among aftermarket customers. It provides

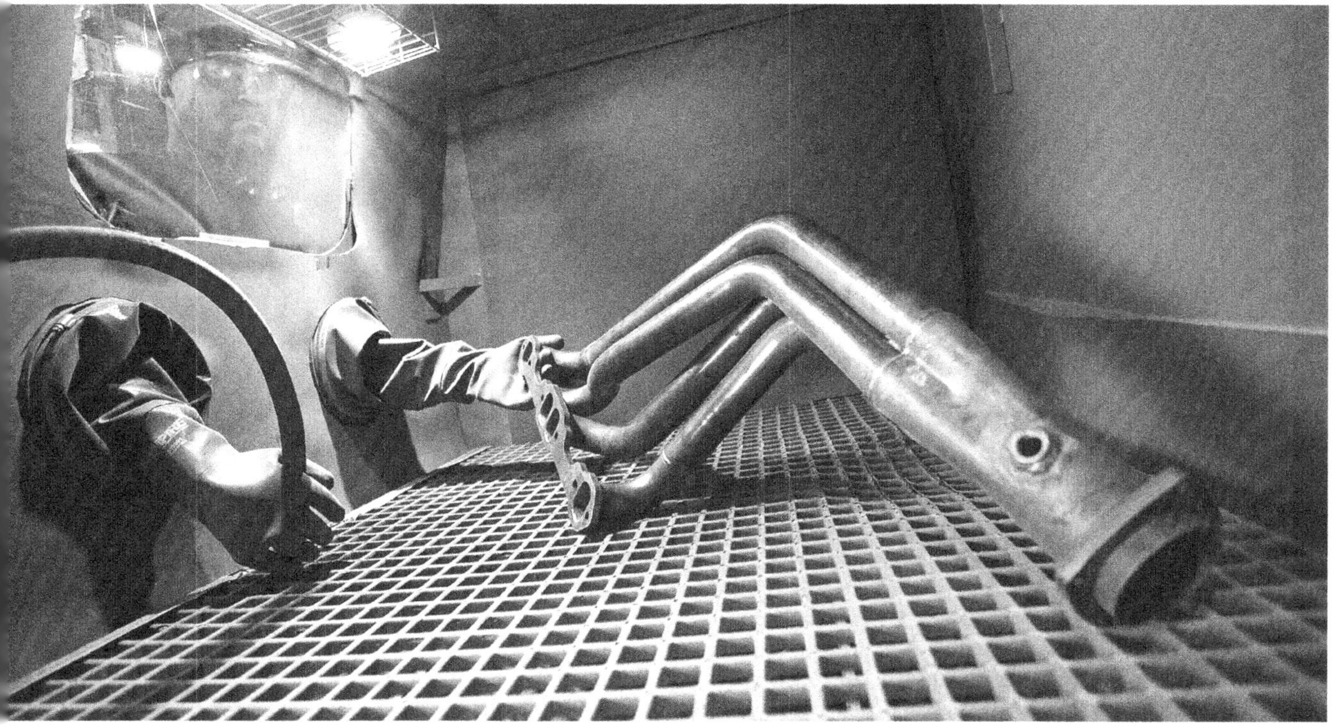

exceptional heat management, prolonging the life of parts while enhancing performance. Available in classic polish and multiple colors, Jet-Hot Extreme 1300 will hold its luster year after year and is covered by a lifetime warranty.

Jet-Hot Extreme 2000 is often associated with turbo applications and can withstand temperature environments of 1,300-2,000 degrees F. Jet-Hot Extreme 2000 is available in black, gray, and several other colors, and has a gritty, rough texture.

Jet-Hot Extreme 2500 series eats these for breakfast, with cayenne pepper on top. So if you need serious heat management with major corrosion protection, bookmark this page.

Jet-Hot Maintanance Tips

When dyno testing or breaking in a new motor or valvetrain it is very important to ensure adequate cooling of heat-quenched coated exhaust components. Stagnant air or no vehicle movement combined with high rpm settings can result in component temperatures exceeding the limitations of the aluminum in the coating matrix. To eliminate this problem, we suggest a large fan be used to dissipate and transfer heat.

To maintain a beautiful finish and prevent corrosion for years to come, periodic maintenance of the coating is required. We recommend periodic cleaning of your parts to remove road grime, salt, and oil from the coatings and any non-abrasive aluminum polish for heavily oxidized or stained coatings.

Stains such as oils, transmission fluid, and antifreeze can be easily removed with fine steel wool or Scotch-Brite (red) and any non-abrasive aluminum polish. After removing the stains, polish to a high luster with the aluminum polish and a soft cloth.

Colors other than polished aluminum require special handling when installing. These are all top-coated systems and the pigments can be very soft in nature. Although tougher than paints, they should be handled with the same care. Remember, all of Jet-Hot's top-coated systems protect against corrosion and thermal fatigue.

Unequal fuel distribution or advanced spark timing can result in extreme EGT's (exhaust gas temperatures), which will exceed the temperature limitations. Make certain all cylinders are functioning properly. Static-time your engine prior to starting it up for break-in. Please call or email Jet-Hot's technical department with any questions at (800) 432-3379 or sales@jet-hot.com.

Do not use any kind of caustic solutions to clean your Jet-Hot–coated components. A variety of chemicals are being used on racetracks and highways today, so make it a habit to occasionally wash off your headers. ⋆

CHAPTER ELEVEN

Amped Up

Wiring, Gauges, Sounds, and Keeping Cool

By Ron Ceridono | Photography by Brian Brennan

Back in 1959 life was pretty simple. *Bonanza, American Bandstand, The Twilight Zone* and *The Adventures of Ozzy and Harriet* were a few of the hit television shows on the three networks that came over a roof-mounted antenna or a set of rabbit ears. There wasn't a home computer or video game to be found and a new Chevy Impala had a base price of $2,717 plus options. Of course the option list was pretty limited compared to what we expect today—a push-button radio added $87 to the cost, A/C with heater was $468, power windows and front seat each jacked the price $102, and a heavy-duty 35-amp generator to power it all was $8. Given what we are used to today if we tried to stuff Jerry Dixey, our Road Tour leader, into a car that was sparsely equipped and send him off to travel 25,000 miles in a summer we'd have to blackmail our tour leader with pictures from Woodstock to get him to drive to the end of his driveway.

Like all the AMSOIL/*STREET RODDER* Road Tour cars our Impala is being built to extremely high standards. Hot Rods by Dean has pulled out all the stops to make this one incredible car and as you might expect that means it's equipped with state-of-the-art electronics, convenience accessories, gadgets, and gizmos. To supply all those current dependent devices, an up-to-date electrical system is required, so we chose a Painless Performance Products customizable

Hot Rods by Dean shot the dashboard of the AMSOIL/*STREET RODDER* Road Tour '59 Chevy with the same PPG Ferrari Red that covers the body. They then installed the custom Dakota Digital VHX gauges and the refurbished trim from Sherm's Custom Plating.

To replace the tattered original wiring in our Impala HRBD installed a Painless Performance Products harness. To make things easy the wires are color-coded and labeled.

For easy access the fuse box was mounted to the firewall to the left of the Flaming River steering column. Painless bundles the wires according to their destination to make routing clean and simple.

wiring harness (PN 10202) for a dash-mounted ignition switch.

While the new 18-circuit Painless harness will provide power for all the Road Tour car's electrical needs, there are a host of other benefits as well. Often overlooked is the superiority of the modern connectors, the quality of the wire, and that it is color-coded and labeled for ease of installation

Installing a Painless Performance Products harness can broken down into four simple steps:

1. Mount the fuse block
2. Route the wires
3. Cut off the excess wire
4. Terminate the wires

To make installation easier the wires coming from the fuse block are broken up into three groups:

Engine/Headlight Group: includes high beam, low beam, park, right turn, left turn, electric fan, horn, starter solenoid and battery feed, alternator and alternator exciter wire, distributor, water temperature, oil pressure, and air conditioning.

Dash Group: includes wires to connect gauges, indicator lights, and switches to their proper sources. Trunk accessory, door locks, power windows, and electric fuel pump are included in the 18-circuit kit.

Rear Light Group: includes taillights, dome lights, left and right turn signals, brake lights, and fuel sender.

A Look at Electrical System Components and a Few Tips

By Dennis Overholser/Painless Performance

Circuit Breakers: Circuit breakers come in many different amperage capacities and are used in circuits the have high loads and voltage spikes. An electric fan circuit is a good example of a circuit that a circuit breaker should protect.

Wire Stranding: Did you know that in theory current flows around the strands of wire? This means that the more strands of wire in a wire the more current that wire can carry. When doing wiring of a circuit, be sure the wire is high quality and has maximum strands.

Ignition Switches: Most aftermarket ignition switches are limited to about 60 amps of current-carrying capacity. Always use caution when using a universal switch that it is not overloaded by it powering the electric fan circuit or the air conditioning circuit without a relay.

LED Flashers: LED flashers are required in many vehicles because of the installation of LED tail and front turn signal assemblies. LEDs do not pull enough current to trip the bi-metallic strip in a conventional flasher like the 1187 styles bulbs. LED flashers have built-in resistance to solve the problem.

Good Grounds: Grounding of your ride is essential. Always connect the battery ground to the engine and tie the engine to the frame as well as the body with heavy ground straps or cable. If your dash gauges fluctuate when you turn on the lights, there is a grounding problem.

Computers and Welding: Never weld on the chassis or other parts of your ride without disconnecting the battery and also the computer, if so equipped. Electrical

continued on next page

The various groups are routed to the areas where they will connect and secured to prevent damage. In some cases a wire may not be used. For instance if the harness has a wire for an electric fuel pump and it's not used, it can be left in the harness in case a change in made at a later date. Thanks to color coding and labeling, any unused wires will be easy to identify if and when it is needed.

Painless offers a variety of kits and components to make wiring easy. The proper fuses have been pre-installed in the fuse block; all wires are color-coded and labeled. In addition Painless kits have "accessory" terminals at the front of the fuse block that are un-fused but may be constantly hot or "switched" hot but by adding an in-line fuse or circuit breaker. For safety and reliability all wire is 600 V, 125 degrees C, TXL. Standard automotive wire is GPT, 300 V, 80 degrees C, with PVC insulation.

INSTRUMENTS

As the name implies, Dakota Digital specializes in digital instrumentation, manufactured in their facility on the outskirts of Sioux Falls, South Dakota.

There are several possible scenarios you'll encounter when outfitting your hot rod with new instrumentation. In short, there are a few different outputs, or sources, and a few different inputs to the gauge.

The oldest and simplest scenario is a mechanical speedometer cable the threads or clips on to the transmission, and in turn

Josh, one of the engineers at Dakota Digital, works our gauge design. The Road Tour '59 will receive one of the first systems with a new font and revised, high-definition tick mark arrangement.

Here's our instrument cluster from the rear. Note the ribbon connectors; they simply plug into the VHX control box—they really clean up the mess underdash wiring can be.

A Look at Electrical System Components and a Few Tips

Continued from page 00.

surges from welding can cause internal circuitry failure in a computer and could cause gases from the battery to explode.

Alternator Output: When choosing an alternator for your ride, always ask what the output is at idle. We all idle around the fairgrounds and the output at idle is more important than the maximum output rating. Be sure to only purchase one that has a minimum of 70 amps at idle.

Harness Routing: When installing a chassis harness and a fuel injection harness, never route them through the same firewall hole. The radio frequency noise from the chassis harness could cause the signals from the sensors to the computer to get distorted and cause rough engine operation.

Battery Charging: When recharging the battery in your ride with a battery charger, always hook up the positive lead first to the battery and the negative lead to a good ground on the engine or chassis. This will prevent a spark close to the battery, which could cause an explosion of the batter gases.

Test Light/Voltmeter: When testing circuits, a test light or voltmeter can be used to check circuits in the main chassis harness but only a voltmeter should be used when testing circuits in the fuel injection system. This will prevent unexpected loads on sensitive circuitry in the computer.

Electric Fans: On average, an electric fan will require 1 amp per inch of overall diameter to operate at normal speed. For instance a 12-inch fan will require 12 amps to operate but also note that twice the amperage is required during clips or threads onto the back of the speedometer. Adjustment to compensate for tire size and gearing changes is accomplished by swapping the driven-gear in the transmission.

The next most common situation we encounter is an older transmission ready to accept a mechanical cable matched with an electronic speedometer. Instead of a cable spinning the needle in the gauge, we'll use a speed sensor, or pulse generator, connected to the transmission. The generator reads the mechanical rotation of the driven gear in the transmission and converts the movement to an electrical signal that is in turn read by the gauge. Adjustment is typically accomplished through the gauge; most quality aftermarket speedometers have a calibration process. If your speedometer doesn't have a calibration procedure, or the speed reading is too far out of range, an additional inline interface such as the Dakota Digital SGI-5 is usually required.

Starting in the '80s, transmissions moved away from mechanical cables and started to have internal speed sensors. As you can imagine, this can save a few steps and simplify your build if you're using aftermarket, electronic instrumentation. Typically these electronic-output transmissions require some sort of computer or controller to operate properly; in this case, the internally generated Vehicle Speed Signal (VSS) information must be routed directly to the controller as the factory intended. Typically

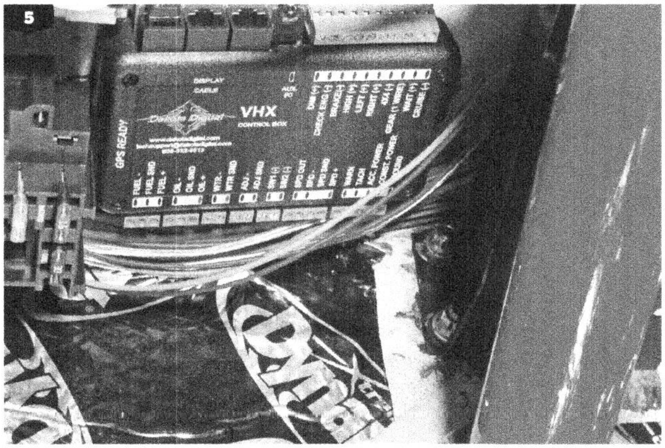

This is the Dakota Digital VHX control box. Along with supplying gauge input it facilitates warning/turn signal lights, gear position indicator (with the proper sender) cruise control (with additional components) as it is GPS ready.

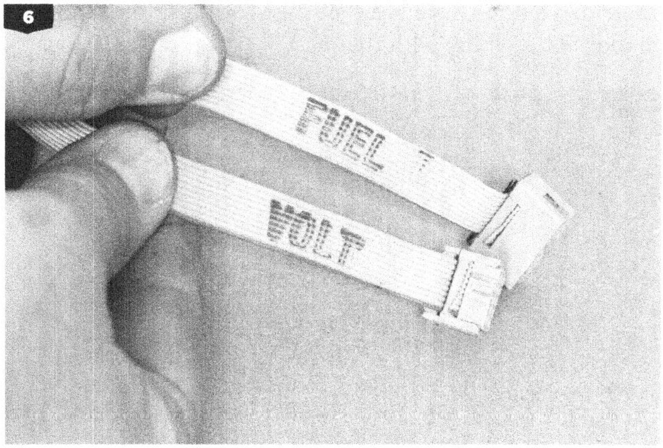

Each ribbon connector is marked appropriately. Wiring the gauges is simply a plug-and-play operation. Part of the Hughes Performance 4L80E transmission package is their transmission controller. It was mounted to the firewall on four studs.

initial startup. Be sure the circuit protection can handle the startup load as well as the fan relay.

Hot/No Start Problems: GM large starters are well known to not activate when they get hot from driving. This is due to the solenoid plunger expanding due to the heat and sticking in its housing. A relay between the solenoid activation terminal and the battery source, to flow more current, will almost always remedy the problem.

Soldering vs. Crimping: As the old saying goes "soldering is great on TV sets." Controversy is long standing on this subject. The general rule of thumb is that if you are proficient in soldering and the wires being soldered will never move, solder them. In automobiles, wires are mostly moving and vibrating, which can cause a solder joint to break the wire next to the joint. Crimping is preferred.

Solid State Devices: The term solid-state means that an electrical device has no moving parts other than electrons. Some of the common solid-state devices found in trucks are:

Diodes: These are one-way electrical valves; current can flow one way but not the other. Diodes are found in a number of places, most notably in alternators; they help transform alternating current to direct current.

Light-Emitting Diodes: Becoming more common all the time, LEDs have a small lens built in. The flow of electrons generates energy released in the form light. Unlike a conventional bulb, no heat is created.

Transistors: These are solid-state switch used in applications like electronic ignitions to turn a circuit on and off, or in a stereo to strengthen a radio signal.

Resistors: Used to limit current flow, resistors can have a fixed value or they can be adjustable (also called a potentiometer).

Capacitors: A capacitor can store and electric charge. The condenser in a point style ignition is actually a capacitor. Some electronic ignition systems use a capacitor to store an electrical charge that is released all at once.

the trans controller afforded a VSS output to the stock cluster, if you have this, use it for your gauge. If you cannot locate a VSS source, a speedometer interface like the SGI-5 can be spliced inline and successfully split the speed signal, allowing it to be shared by two devices (the speedometer and transmission controller).

If you have a modern transmission with no provisions for a mechanical cable, but have your heart set on an old mechanical speedometer, there's a solution for that, too! Electronic cable drive units have been on the market for a while, with the most recent entry to the market being the ECD-100 from Dakota Digital. This powered unit can spin a mechanical speedometer cable based on VSS from an ECM or transmission output. The ECD-100 unit handles calibration, and a 36-inch cable is included.

The latest development in speed information for hot rods is GPS technology. When a speedometer is connected to a GPS sensor, the speed information is no longer generated by the vehicle itself, but by satellites tracking its movement. The benefits include simplified installation, and in some scenarios with race transmissions, there are no provisions for a VSS connection so GPS is the simplest choice. The downside of GPS technology is the initial cost and consistency. Since the GPS unit in the car is getting information from outer space, there will be a short period of time required to acquire adequate satellite signals. Much like satellite radio, reception isn't always perfect and speedometer performance can be affected by physical obstructions, such as bridges, heavy tree cover, and buildings. Many people assume that there is no calibration or setup required with a GPS-based speedometer system, but this isn't always true. You'll still need

7

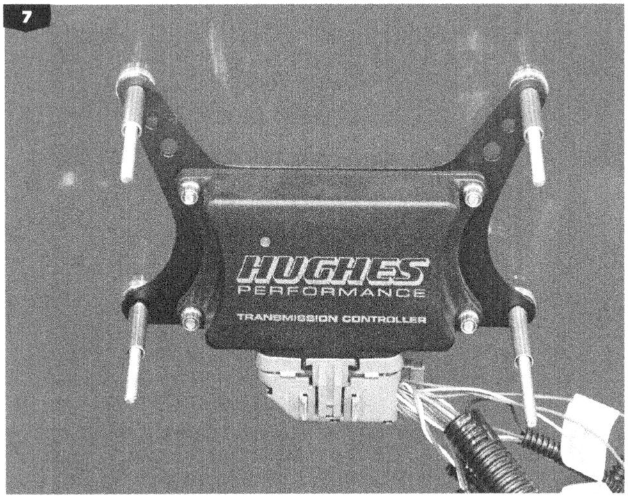

Along with the transmission controller, HRBD fashioned mounts for the FAST EFI and EZ LS ignition controller.

8

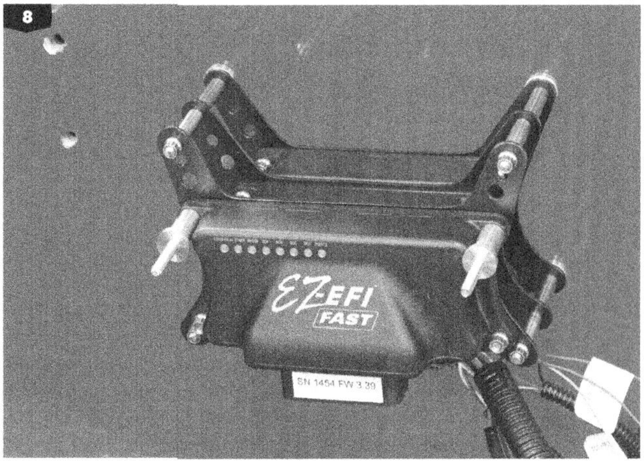

To protect and hide the control units HRBD fashioned an easily removable custom cover.

9

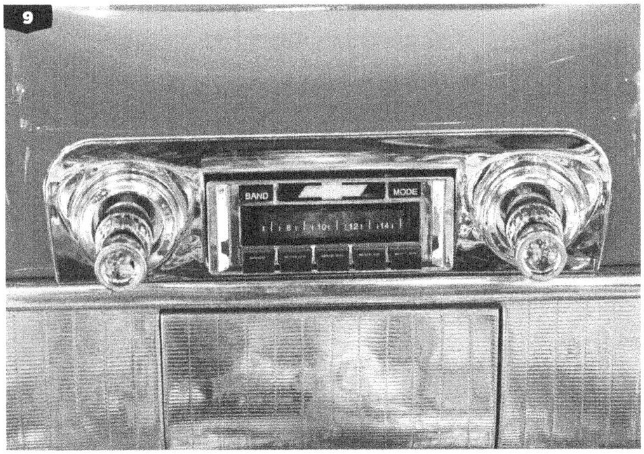

Tunes for the Road Tour Chevy are supplied by a complete Custom Autosound system. The dash unit fits in the stock location.

10

To make our Impala into a rolling concert hall, Custom Autosound supplied a compact bass enclosure with an integrated amplifier.

to wire the GPS unit into the car, connect it to the gauge, and in many cases, ensure it's properly using the GPS information supplied.

Tachometer

Most hot rods have a tachometer; installing a rev counter is a rite of passage for many budding rodders. Connecting a tach is a basic procedure, but can vary depending on your engine. Starting with the most basic, for those still running points, or have an external ignition coil, the majority of aftermarket tachometers are looking for a signal from the negative side of said coil. HEI distributors offer a dedicated TACH terminal to connect your new gauge, and most CD-type ignition boxes have the tach signal wire singled out. With the landslide of late-model engine transplants taking place, things can get a bit murky when wiring in your new tach.

If you're among the lucky ones, your ECM wiring harness offers up a tach-signal output wire. If not, you'll have to get a bit more creative; tapping into one coil a multi-coil system can work, provided your tach allows adjustment for cylinder count. You may need to employ a tach-pickup unit from MSD along with a signal modifier, such as the SGI-8 from Dakota Digital. If your hot rod is using an OBD-II computer system, much of the information you seek can be drawn from the ECM. Dakota Digital offers the STA-1000 interface that plugs directly into the diagnostic port on an OBD-II equipped vehicle, and can output speed and tachometer information for most aftermarket instruments.

Oil Pressure

When it comes to measuring the lifeblood of your engine, tracking oil pressure is relatively straightforward. There are two methods to reading oil pressure: mechanically and electronically. With a mechanical gauge, there is a capillary tube, which runs from the engine all the way to the gauge. Many feel this is the most accurate way to measure pressure, but the downside is the capillary tube connection; with many cheap gauges, the connection is made with a thin plastic tube that can easily become kinked or worse, cracked. Better instruments use metal tubes, but are still a bear to route and work with.

Many modern instruments sets make use of electronic oil pressure gauges, which utilize a sensor threaded into the engine that converts pressure into voltage or resistance; this information is relayed via wires to the instrument. Because manufacturers design their instruments around a certain sending unit, it is extremely important to use the correct sending unit, as prescribed by the gauge manufacturer. Never assume an existing sensor, or a factory sensor, will be compatible with an aftermarket gauge; we've seen gauges read very inaccurately, read backward, or simply offer no reading at all when paired with the wrong sensor. The latest advancement in instrumentation technology is solid-state sending units, which offer the utmost in accuracy and consistency.

Water Temperature

Water temperature is a critical reading for any engine, so an accurate gauge is a must have. Much like oil pressure, there are two methods to taking your temperature: mechanical and electronic. A mechanical water temperature gauge uses a brass bulb in the engine with a long capillary tube reaching to the gauge. Hard to install, prone to kinks, and easily influenced by radiant heat, the mechanical water

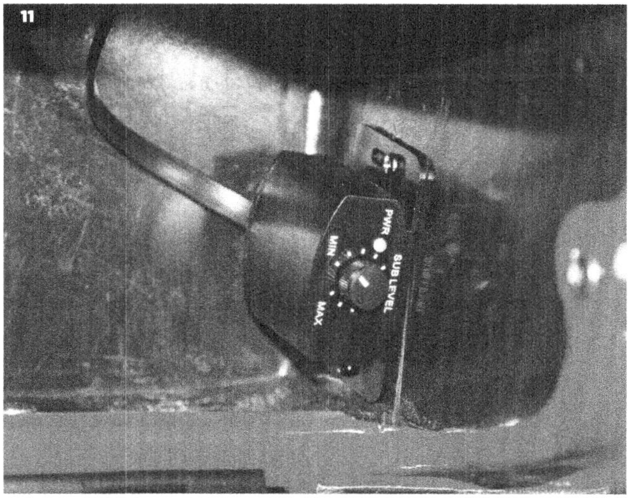

The control for the sound system's subwoofer was hidden in the glovebox (part of the Vintage Air Direct Fit kit).

In classic configuration a pair of Custom Autosound 6x9-inch speakers were installed in the package tray.

temperature gauge is found most often in vintage restorations.

Electronic temperature sensors thread into a water jacket and convert temperature information into resistance, which is sent to the gauge with one or two wires. This takes much of the hassle out of installation and routing, and increases accuracy and consistency. The catch is ensuring you have the sending unit specified by the gauge manufacturer; many sending units look alike, but a factory temp sender is not compatible with aftermarket instrumentation. Tech tip: when installing your new temperature sensor, be mindful of heat sources that may skew your temperature readings; many OEM senders are installed in the cylinder head and are calibrated to compensate for heat off the exhaust. Most aftermarket companies recommend locating the temperature sender in the intake manifold, away from exhaust heat.

Voltage

Being able to see what your electrical system is doing allows you to see any charging issues before they leave you stranded. A quality volt gauge will report how many volts are coursing through your electrical system through the fuse panel. Connection for a volt gauge is simply a power wire; a volt gauge does not directly monitor the battery, or the alternator output, but rather the system as a whole. Much like in your home, an electrician will probe an outlet or the breaker box, but not the transformer outside. We want to see how the power is moving within the system.

Fuel Level

Keeping tabs on how much fuel you have is as important as it is simple. Typically, a float-style sending unit is mounted in the gas tank and sends a resistance value to the gauge via a single wire. There are a few floatless senders on the market, but connectivity to the gauge remains similarly straightforward. First and foremost, ensure your gauge is compatible with the sending unit you have; advanced aftermarket instruments can be calibrated or adjusted to accommodate an existing sender. In any case, follow the gauge manufacturer's instructions to the letter to ensure you don't run your tank dry on the road.

Recently Dakota Digital introduced a new line of analog gauges. The new VHX instruments feature fully lit needles, backlit faces, and highly visible LCD message centers.

Hanging on the front of the 427-inch LS1 built by Ron Shaver Racing Engines is a Vintage Air FrontRunner drive system. It's clean and compact and mounts the A/C condenser, alternator, and power steering pump and drives them all with serpentine belt.

VHX

What Is VHX?

VHX is a loose acronym for Vehicle Hybrid Instrument Systems, however the name itself does not really define the VHX lineup. This revolutionary product line will quickly evolve beyond the applications listed to cover several popular hot rod, muscle car, street rod, and custom vehicle applications. From direct-fit applications to a universally mounted approach, the benefits are truly limitless. Fully lit needles, backlit faces, and highly visible LCD message centers are a few of the things that you can see which set the VHX Series apart from yesterday's approach to aftermarket instrumentation. A few of the features which you don't normally see are the micro-controlled precision stepper motors, solid state sensors for unparalleled accuracy and user-customizable display feedback, all supported by an unrivaled limited lifetime warranty which has become the standard for Dakota Digital instrumentation systems. Completely engineered and manufactured at our facility in Sioux Falls, South Dakota, the VHX Series has completely revolutionized the instrumentation industry, making your decision quite simple.

THE VHX SYSTEM

Displays included with the VHX system:

- Speedometer: 0-120/140/160 mph analog depending upon application, 0-255 mph/kph digitally in message center.
- Tachometer: 0-8,000 analog, 0-9,990 digitally in message center.
- Oil Pressure: 0-80-psi analog and digitally.
- Water Temp: 140-220 degrees F analog, 0-300 degrees F/ 0-150 degrees C digitally in message center.
- Voltmeter: 0-17VDC analog and digitally.
- Fuel Level: Empty-Full analog. 0-99 percent digitally in message center.

Also shown in the digital message center:

- Odometer
- Trip Odometer (A&B)
- Resettable Service Odometer
- Clock
- Hour-meter
- 0-to-60 mph (0-100 kph) timer
- quarter-mile timer
- quarter-mile speed display
- High-speed recall
- High-rpm recall

Various Indicators can also be displayed such as:

- Turn Signals
- High Beam
- Check Engine
- Parking Brake
- Cruise Control
- Automatic Transmission gear indicator is also included, which will display the full name (Park, Overdrive, and so on) of the current gear. Utilizing the gear position indicator will require the use of a Dakota Digital GSS-2000 sending unit (sold separately).

Special Features

Dakota Digital instrument systems also include a demonstration mode, which will cycle the instrument readouts while the vehicle is standing still. This is especially popular while the vehicle is on display at a show, event, or simply displaying the instrument system.

Metric conversions will cycle between mph or kph in the digital message center with the press of a button. Full metric analog instrument systems (kph and celsius) are not currently available.

Necessary sensors included for oil pressure (1/8-inch NPT), Water Temp (1/8-inch with bushings for 1/4-, 3/8-, 1/2-inch NPT as well 12mm and 16mm for GM & LS applications) and speedometer pulse generator.

Fuel level gauge can read either OEM sensor, or any aftermarket style currently available. A universal fuel level sender, SEN-06-1, can be purchased from Dakota Digital if necessary.

Fully programmable speedometer for various tire sizes, rearend gearing combinations, and so on.

Built-in illumination is activated as headlights/parking lights are activated. The DIM-1 can be purchased from Dakota Digital to increase or decrease the brightness of the display. The DIM-1 is used in conjunction with your headlight switch

Fully adjustable warning points alert the user of potential vehicle problems in the digital message center.

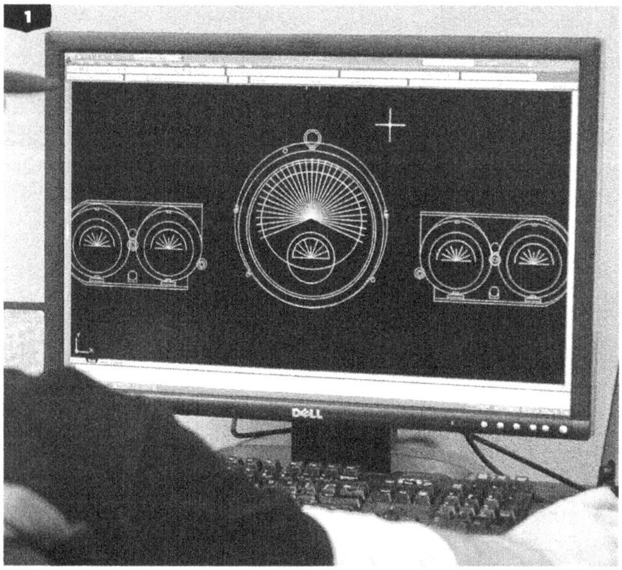

With the mounting points, viewing windows, and basic structure of the system laid out using CAD software, Josh, a mechanical engineer, creates a polar array to determine the placement of tick marks on the gauge face.

Additional aspects of the instrument system are added to the rendering; in this photo you can see the indicators, annunciators, openings for the LCD message centers, as well as the gauge numbers.

3

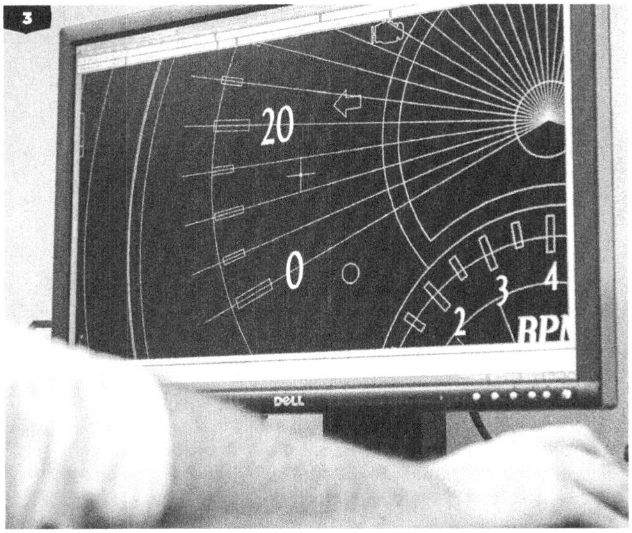

Consistent spacing is crucial to for accurate instrumentation, as well as a visual pleasing finished product.

4

The design aspect is nearly completed, and the green guidelines are removed, allowing the system as a whole to be scrutinized by the entire design team. This is when last-minute tweaks are made; the custom *STREET RODDER* and Impala logos will be added in the next step.

5

Using Photoshop, Josh adds life to the rendering, foreshadowing the finished product. Additional artwork has been added to this overlay, making the instrument system for the Road Tour '59 truly one of a kind.

6

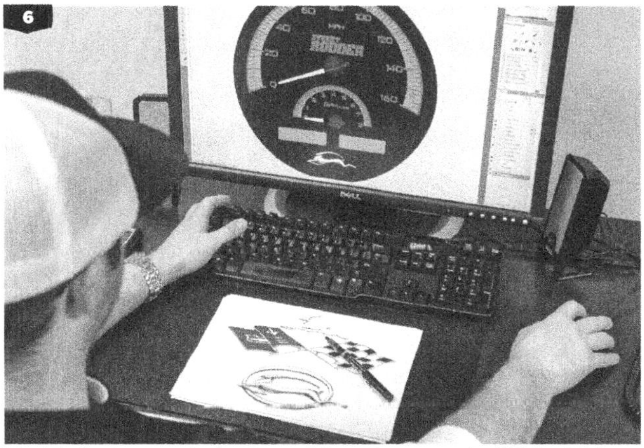

The Photoshop rendering is used for reference only. Think of this as the final proof of the design process.

7

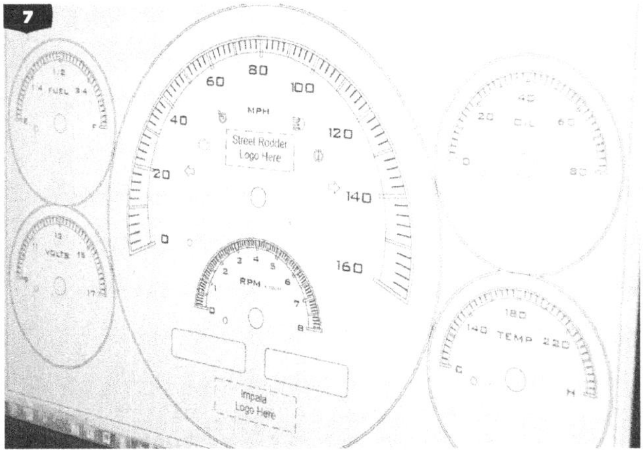

The raw CAD file is imported into another program, and different layers are added to flesh out the intricate details in the overlay.

8

The Road Tour '59 will receive one of the first systems with a new font and revised, high-definition tick mark arrangement. Visible in this photo are the four "gunsights" used by the optical laser that cuts the overlays.

9

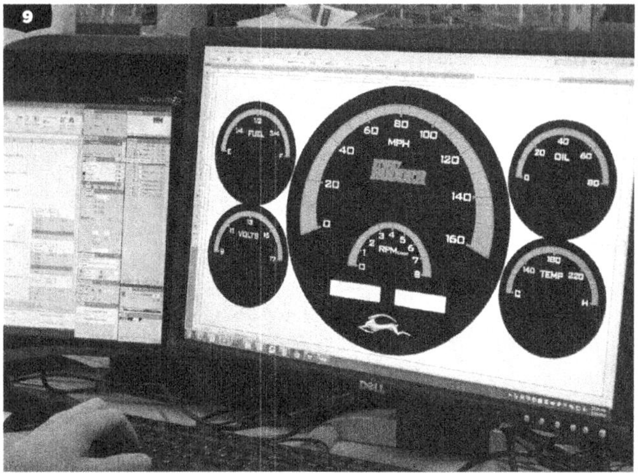

With colors added and the "leaping deer" emblem, the overlay really comes to life.

10

Fine-tuning an overlay design is an in-depth process that requires focus and great attention to detail. The UV printer that creates the overlays offers extremely high resolution, so accuracy is key.

11

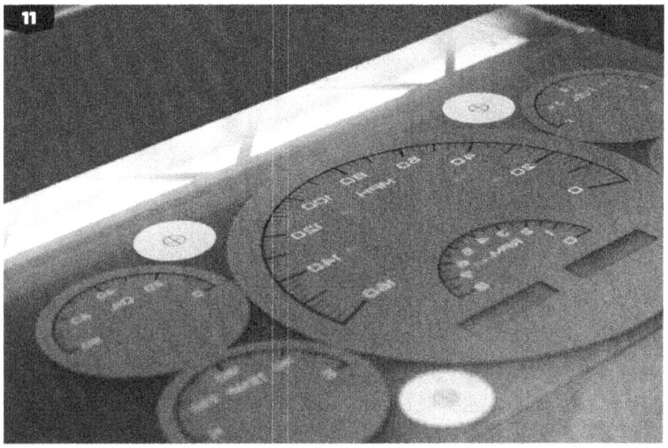

The overlays are printed using an intricate, multi-step process. Many layers of ink are stacked in a specific order to create opaque and translucent areas; thanks to the UV-cured ink, Dakota Digital is able to hide the turn signal, high beam, check engine, along with a few other indicators right on the gauge face.

12

The beauty of the hidden indicators is, should you have no use for a cruise control or check engine light, they'll never light up and you won't know they're there, but are available for the next guy. One product serves many different needs.

13

A freshly printed overlay awaiting a turn under the laser. At this angle, the indicators are visible.

14

The optical laser has its own software, making the process fairly straightforward. An "eye" focuses in on the printed alignment marks to ensure a precision cut.

Many times faster and more accurate than an X-acto knife, the laser makes quick work of the Road Tour '59 overlay.

While the overlay is prepared, work begins on other components of the instrument system. Here, the visible edges of the diffuser are blacked out. These pieces allow light to be placed exactly where it is needed in the overlay.

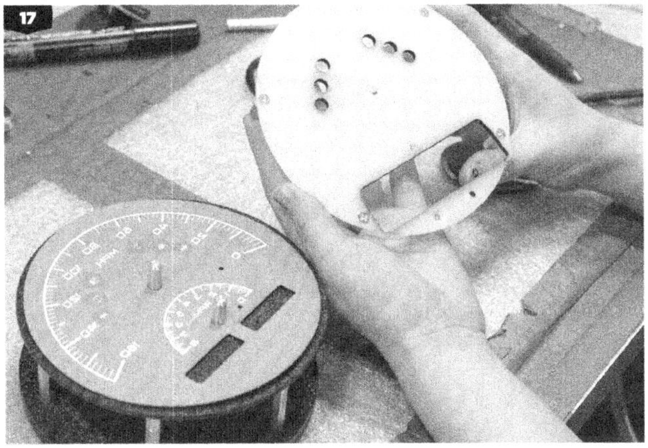

In this photo, you can see the laser-cut holes in the diffuser that allow the LED lighting to be guided to each specific indicator in the overlay.

Stopper pins for each needle are pressed into the diffuser once the overlay is in place. These pins are used by the stepper motors to "zero" the gauge upon startup and ensure an accurate reading each time you start your hot rod.

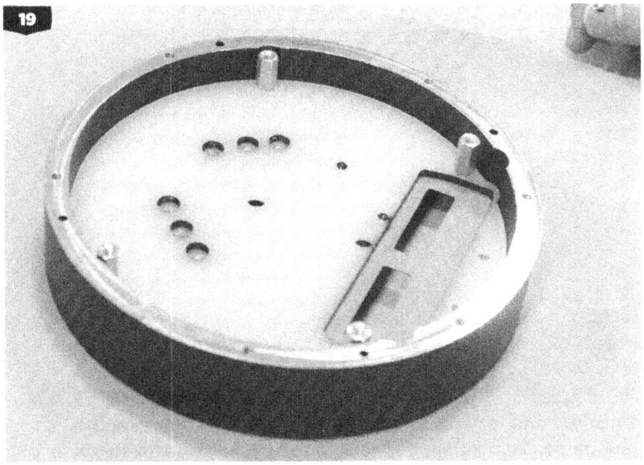

With the powdercoated, billet aluminum housing, overlay and diffusers mated together, it's starting to look like something!

In yet another area, the circuit boards are populated with capacitors, resistors, stepper motors, and LCD displays. VHX instrumentation brings the instrumentation into the 21st century.

21

Small ribbon cables connect the satellite units to the "parent" speedometer unit, while a single CAT5 networking cable links the display unit to the control box. Say goodbye to the plate of pasta behind your dash.

23

Numerous fixtures are used to assemble each instrument system to ensure exact tolerances are maintained.

25

The finished product is placed on what is referred to as the burn-in rack. Burn-in is the final step in testing that each instrument system undergoes; systems are cycled on and off for 24 hours to ensure proper operation of each stepper motor, LED, and LCD message center.

22

Things are really taking shape with the electronics in place. Before the needles are installed, we can see the light guides on the stepper motors that allow LED lighting to pass through the motor and into the needle. Thanks to this small detail, the needle of each gauge will glow along with the backlighting. The LCD displays allow unlimited additional readings to be displayed without cluttering the interior.

24

Pressing the needles onto the motor shafts is one of the last steps before testing. A CNC-cut, scratch-resistant polycarbonate lens is screwed to the billet housing and is the last step in making a sealed unit, free of dust.

26

Once the unit is run through its paces one final time and the unit passes inspection, it is sent to the packaging and shipping department. A protective film is seen over the acrylic lens to ensure a blemish-free final product is shipped to the customer.

TRANSMISSION CONTROLLER

To manage operation of our Hughes Performance 4L80E we used a Hughes Transmission Controller Unit. The TCU (Transmission Control Unit) system is made up of several main components: a TCU, wiring harness, handheld user interface, and an rpm module.

The TCU is water tight when connected to the wiring harness. It can be mounted in the engine compartment or in the vehicle's interior. It is good practice to mount the TCU with the connector facing down. This way, there is less chance of moisture getting into the TCU if it needs to be disconnected in wet conditions.

The TCU should not be mounted close to other electrically "noisy" components. In particular, keep good spacing (try for 2 feet minimum) from ignition components (ignition boxes, coils, distributors, and so on). If mounting in the engine compartment, selecting a location toward the rear will make it easier to route the communications cable to the interior to allow the handheld to be monitored while driving. Optional, longer communications cables are available if needed.

There is an LED on the front face of the TCU—the side with the logo. It will flash if the on-board diagnostics detects a problem (the LED is lit solid when the TCU is powered up and working normally). To take advantage of this feature, the TCU will need to be mounted so that the front face of the TCU is visible. The handheld will also indicate if any faults have been detected.

SOUND SYSTEM

Again this year we've gone to Custom Autosound to provide the tunes on our Impala. Custom Autosound's founder, Carl Sprague, built his first custom sound systems in 1977 for his '63 Corvette. Since then he has gone on to manufacture a complete line of products, including six different radios/CD players, upgrade speakers, and many other related accessories.

Currently CAS offers OEM-fit radios for over 400 different year groups, makes, and models of cars and trucks, '40s-80s, as well as the universal "fits anything" Secretaudio AM/FM CD system.

Custom Autosound kits (modifies) each radio to fit specific vehicles so the radio fits your dash right out of the box.

For our '59 Impala CAS offers the following radio options:

USA-1: For great sound at an affordable price, the USA-1 is a great choice. It offers 200 watts and auxiliary input for iPod, MP3, satellite radio, and so on. Other features include: AM/FM stereo, auto reverse cassette, 200 watts (4x50), auxiliary input, electronic tuning and volume control, preset scan, digital clock, best station memory, 30 presets 12AM-18FM), 4-channel RCA preouts, four-way fader, separate bass and treble, fast forward and rewind, power antenna lead, chrome face also available in black face USA-01BK.

USA-230: Another option is the USA-230. This new model radio has the same cosmetics as the USA-630. Without the USB/MP3 or CD changer control. However, if you have a portable MP3 player or satellite radio, just plug it in to the Aux input and listen to it through this radio. This radio features: AM/FM stereo, 200 watts (4x50), auxiliary input (RCA rear) RCA to 1/8-pin jack included, digital clock, electronic tuner, 16 presets (4AM-12FM), four-channel RCA preouts, four-way fader, and power antenna lead.

USA-630: Custom Autosound's USA-630 has large OEM-style push buttons that give a perfect look in a classic dash. Although it looks classic, it has the latest modern features that include an iPod doc. Navigation for the iPod includes artist/song/albums/playlists, all from the buttons on the radio, so you can keep the iPod hidden away in your glovebox or console. The USB input is still there on the back of the radio and allows you to plug in a USB flash drive for playback of MP3/WMA song files. The new alphanumeric display shows the song/artist/album titles during playback of flash drive (thumb drive), iPod files, and RDS radio station information. As an option you can get Bluetooth music streaming from your Bluetooth-capable phone or Bluetooth-compatible player with the BLUKIT. The radio can even advance the songs on your phone by using the track up button on the radio. It will also mute the music while you take a hands-free phone call through the radio. With 300 watts peak the radio should have all the power you may need, if you want more there are four preout jacks on the back of the radio for optional amplifiers. The back of the radio also has auxiliary audio input for satellite radio or any portable player you may wish to use. The USA-630 feautres: AM/FM stereo, RDS tuner show shows station/song/artist on LCD display, 300 watts peak, USB port for flash drive MP3/WMA playback, iPod doc and control (control your iPod fom the buttons on the radio), direct six-CD changer control, negative LCD display (black background with white letters/numbers), alphanumeric display for MP3/WMA/iPod files, electronic tuning and volume control, loudness compensation, digital EQ, four-way fader, left-right balance, digital clock, two-channel RCA auxiliary input for satellite (cable included), 12 presets (12 FM/4AM), four-channel RCA preouts, separate bass and treble, USA/Euro Tuner, and power antenna lead.

Dash Speakers

Custom Autosound's dual dash speakers provide stereo sound from factory mono speaker locations. There are 38 different configurations available for most classic cars and trucks. The 1000 series speakers replace "OEM" speakers that are held in place by screws. The 2000 series speakers replace "OEM" speakers that are held in place by a metal bracket. These assemblies are pre-wired to "plug in" instantly to Custom Autosound's radios. You can choose from Custom Autosound speakers or try Kenwood speaker upgrades.

Dual Coil Features

Our new dual voice coil (DVC) speakers will give you the best quality sound possible from your original mono speaker location. These speakers have up to 70 percent more cone area than the conventional dual speaker setup, giving you improved bass response and 140-watt power handling. Both stereo channels connect directly to one speaker. This is the best sound solution for the original speaker location we have found.

Rear Speakers

These three-way 6x9-inch speakers are rated at 200 watts and come with Custom Autosound–exclusive flat metal grille with black finish. They can easily be painted to match your interior or left as is. Full Chrome Speaker Basket, 63mm Ferrite Tweeter and 36mm Neodymium Tweeter, 4 ohm; 16-ounce Magnet, 67Hz to 20 kHz frequency response; plus sensitivity 93db.

UNDERCOVER Speaker Enclosures: Custom Autosound designed and manufactured these speakers specifically for under seat or horizontal wall installations. They could be used on large parcel shelves or wherever! Compact unit has one 5-1/2-inch woofer and piezo tweeter in each enclosure.

UNDERCOVER II Upgrade Speaker Enclosures: The UNDERCOVER II has an 8-inch woofer and two Piezo tweeters in each box. This system can be mounted in many of the same locations as the regular UNDERCOVER. With the 8-inch woofer in UNDERCOVERII you get a much fuller and deeper sound.

Kick Panels

For a really clean installation consider Custom Autosound's new kickpanel speaker assemblies. They sound great, enhance interior appearance, and offer a "no modification" fit. Most models feature 6-1/2-inch coaxial (two-way) speakers with 80-watt power handling capacity All kicks can be upgraded with Pioneer 6-1/2-inch coaxial 120 watt. Kickpanels come in black ABS plastic and are sold with or without speakers (except Chevelle models).

AIR CONDITIONING

Of all the accessories ever added to an automobile, air conditioning may be the best. Anyone who has ever gone cross-country during the summer in a car without it will no doubt agree. Traveling with the windows down, vents open, while still being able to fry an egg on the floorboards takes the fun out of the trip. Today virtually every new car is equipped with an effective heating system and A/C, and thanks to Vintage Air many street rods are too, including the 2014 AMSOIL/*STREET RODDER* Road Tour '59 Chevy. To keep our Impala the right temperature at all times we installed a Vintage Air SureFit system with Gen II technology for '59 and '60 Chevrolets (PN 961056). Along with the ability to heat and cool the apartment-size passenger compartment of our Road Tour car, the installation requires no modifications to the dashboard to install.

The heating and cooling provided by the Vintage Air unit is considerably more sophisticated than what was available in 1959. First and foremost the volume of heat delivered is far greater, temperature control is much more precise, and the defrosters are extremely effective. Contemporary electronic heater valves shut off the flow of hot water when heat is not needed, or wanted.

Cooling the air is a little more complicated than heating it. In simple terms air conditioners make the air cold by removing heat.

Here's how the experts at Vintage Air explain the parts and operation:

The components of an automotive air conditioning system are: the compressor, compressor bracket, evaporator (inside car unit), condenser (outside heat exchanger), hoses and fittings, the drier, and a safety switch.

Now, here's what they do:

The compressor pumps refrigerant, in gas form, into the high-pressure gas discharge line. This gas is loaded with heat it has absorbed from the air flowing over the evaporator coil inside the vehicle.

As the heated, high-pressure gas reaches the condenser that is typically mounted in front of the radiator, the air flowing through it carries off the heat. The refrigerant condenses into a liquid, which pours into the receiver/drier where it falls to the bottom of the receiver. A pick-up tube fits into the receiver/drier almost reaching the bottom of the tank (the open end of the tube is always below the liquid level in the receiver/drier, if the system is fully charged) provides pure liquid to the line between the drier and the expansion valve. The expansion valve is an orifice that varies in size according to the temperature of the evaporator coil. By changing size, it meters the refrigerant according to the demands of the evaporator. When warm, the orifice is largest and gets smaller as the evaporator gets colder. This orifice provides a pressure drop with the resulting drop in refrigerant temperature. The high pressure liquid pours into the larger opening of the evaporator and the low pressure liquid droplets begin to pick up heat, which expands the refrigerant to a low pressure, super-heated gas, which then returns to the compressor to start the cycle all over again.

Vintage Air's new Gen II Technology adds a long list of features to aftermarket air conditioning that was previously found only on modern OEM systems. Thanks to the elimination of cable controls and the integration of electronic servos there are dramatic gains in efficiency. The butterfly doors that direct both warm and cool air can travel farther and seal better thanks to the change. The new design provides superior defrost performance and true bi-Level operation in both heat and A/C modes.

In addition to the elimination of cables, improved coil technology and CAD-designed case provides greater cooling/heating capacity and a more efficient air mixing chamber in the smallest possible overall case size and a new servomotor-operated heater control valve provides temperature blending in all modes of operation.

In addition to all the other advantages of the Vintage Air SureFit system are the really cool (no pun intended) cable converters that allow the original controls to operate the new system. Basically the converters do just that, they convert the mechanical movement of the levers to an electrical signal. Calibration is easy enough—set the maximum travel, then the minimum travel and the system figure out what to do in-between.

To mount the compressor on the LS1 powerhouse we chose a Vintage Air FrontRunner system. This totally engineered package solves the most critical shortcomings of conventional aftermarket accessory brackets, such as premature belt failures caused by mechanical tensioning, excessive belt length runs, awkward component placement, and cylinder head mount-bolt variations.

As Vintage Air explains it, "The exclusive FrontRunner spring tensioner provides continuous SAE-spec static and dynamic belt tensioning for OEM

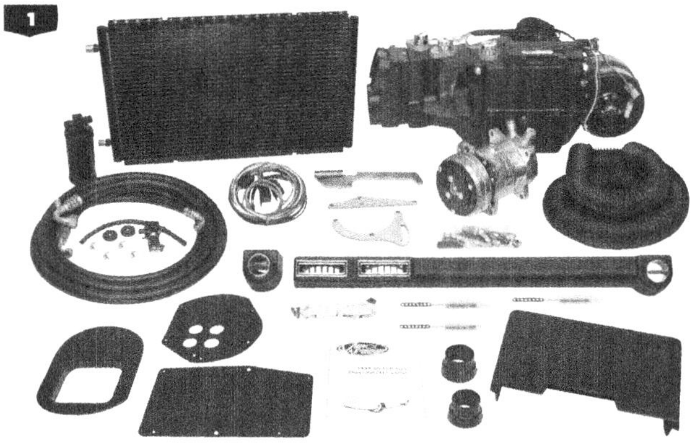

1

Hot Rods by Dean installed a Vintage Air SureFit heater and A/C system for '59-60 Chevrolets in our Impala. The system provides everything needed, even a new glovebox.

2

One of the advancements provided by Gen IV technology are these electronic stepper motors that eliminate manual operation of the butterfly doors, which direct airflow.

3

The Vintage Air unit mounts behind the dashboard. To simplify installations mockup cases are available for many of the systems offered.

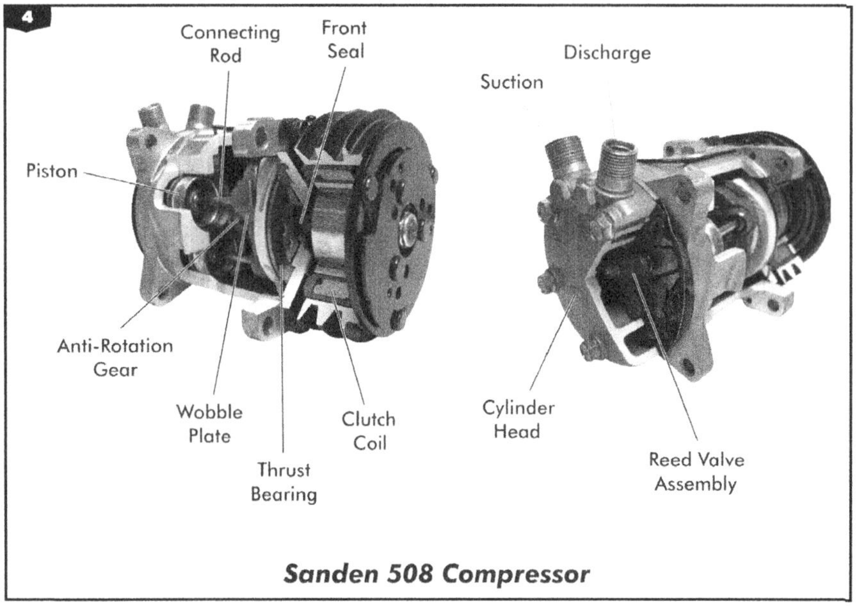

The Sanden compressors used by Vintage Air are compact, smooth running, and reliable. The five connecting rods and pistons are moved by what's called a "wobble plate."

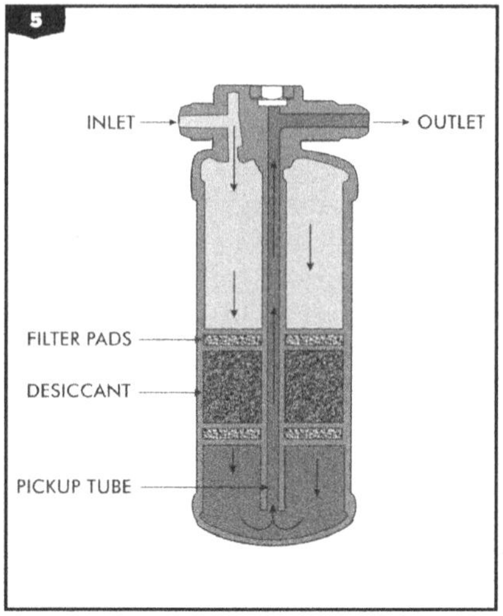

This is a receiver-drier. Its functions are to separate gas and liquid and remove moisture form the system. As liquid refrigerant is picked up from the bottom of the canister, it should always be mounted vertically.

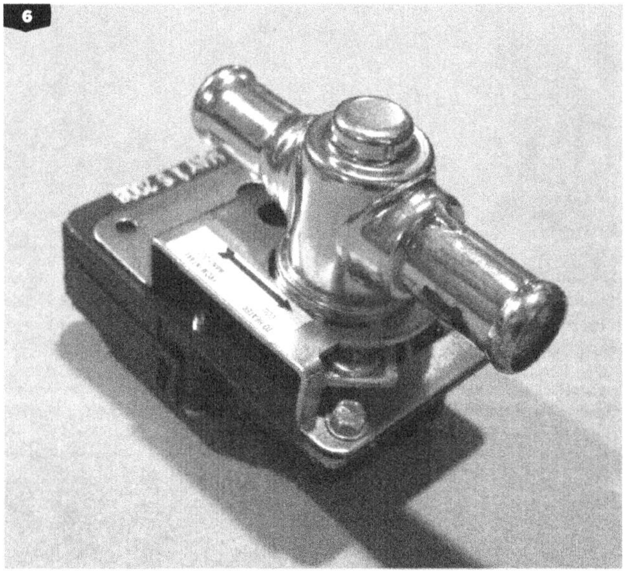

This is an electric hot water valve that eliminates cables and vacuum hoses. It must be oriented correctly—an arrow indicates the direction of water flow.

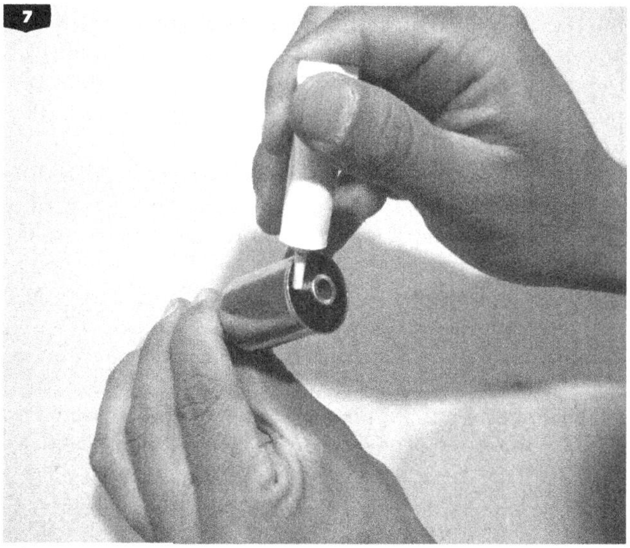

A/C hoses have to be cut to length and the ends crimped on. The ends with small caps are service ports used to evacuate and charge the system. When installing ends on the hose the prescribed lubricant should be used.

quality belt life and the block-mounted design allows any deck height/cylinder head combination as it uses no head or intake mounting bolts like conventional brackets."

Thanks to the truss-architecture design that is CNC machined from T-6061 billet aluminum the FrontRunner is extremely rigid. Pulley sizes are engineered and matched for optimum component speeds and the system is available in machined or gray anodized finish and the precision-machined hard-coated aluminum pulleys reduce weight and wear like steel. Engineered to clear most GM and aftermarket tuned port fuel-injection systems the FrontRunner fits into extremely tight chassis configurations. ★

8

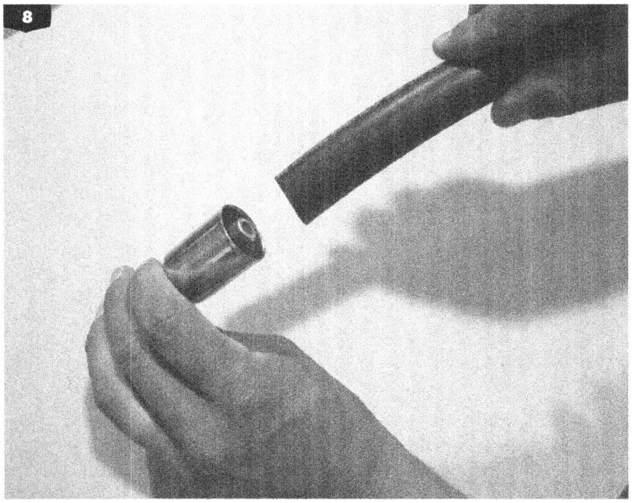

The R134A systems must use barrier hose to prevent the loss of refrigerant.

9

Because of the high pressures involved it is critical that the proper crimping tool is used for the types of ends installed.

10

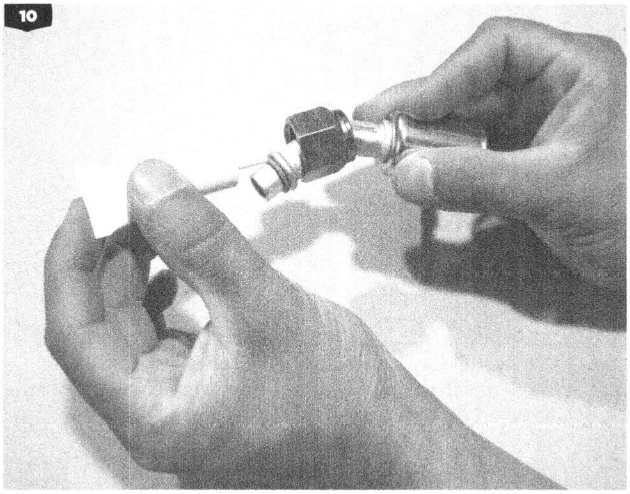

The O-rings used on the fittings must be for R134A; they should be lubricated with the proper oil before installation.

11

An important part of an air-conditioning system is a safety switch. These are extremely important since an A/C system relies on refrigerant to carry lubrication through the system. This is a binary pressure safety switch; it disengages the compressor clutch in case of refrigerant loss.

12

A trinary switch combines low-pressure protection with high system pressure protection that turns on an electric fan at 254 psi.

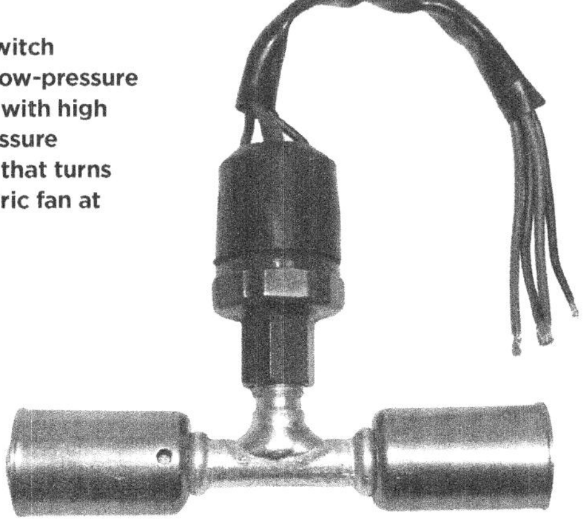

13

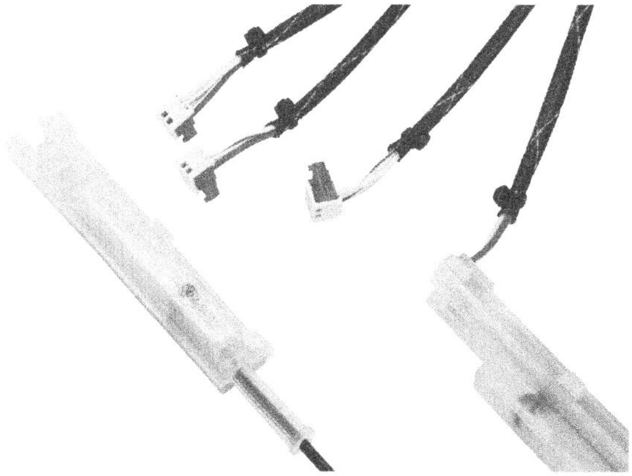

These things are Vintage Air's patented cable converters—they convert the stock cable controls to electronic operation.

14

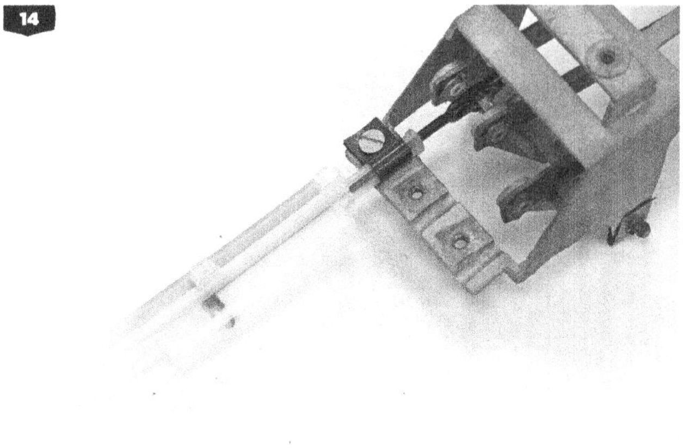

The converters attach via the clamps that once held the cable in place.

15

Here the stock levers have been reinstalled in the housing.

16

The converted controls look like they came from the factory, but the system they operate is far better than what was available in 1959.

17

Once the converters are attached and the control is put back in place it's a simple matter of plugging in the wiring harness from the SureFit heat and A/C unit.

18

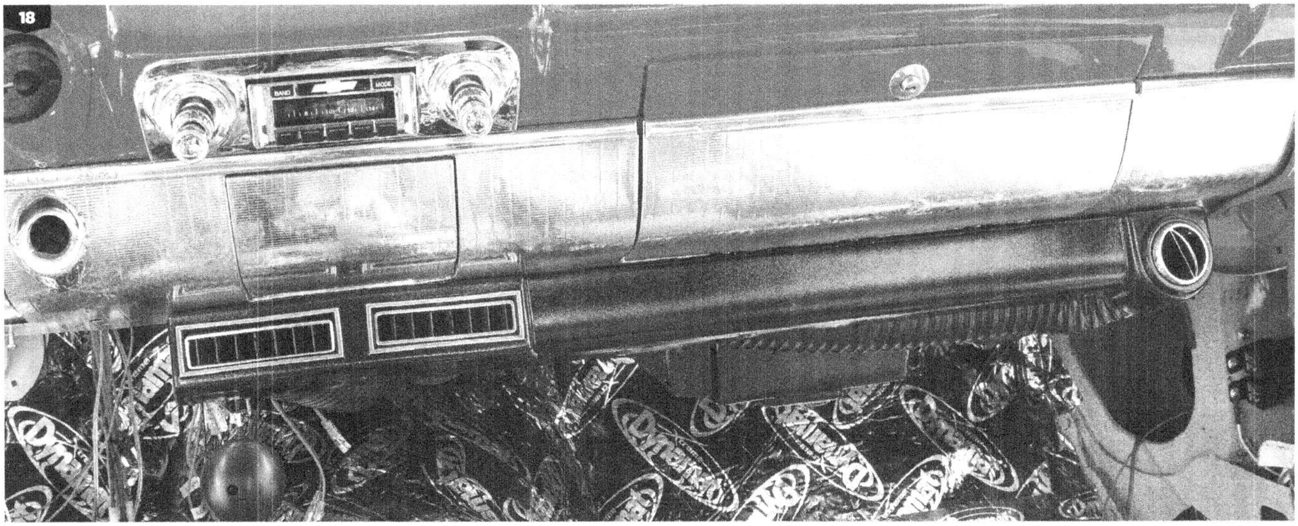

Vintage Air's SureFit system installs without any modifications to the dash. Note the aluminum trim that looks better than new, thanks to Sherm's Custom Plating.

Basic Air Conditioning Components

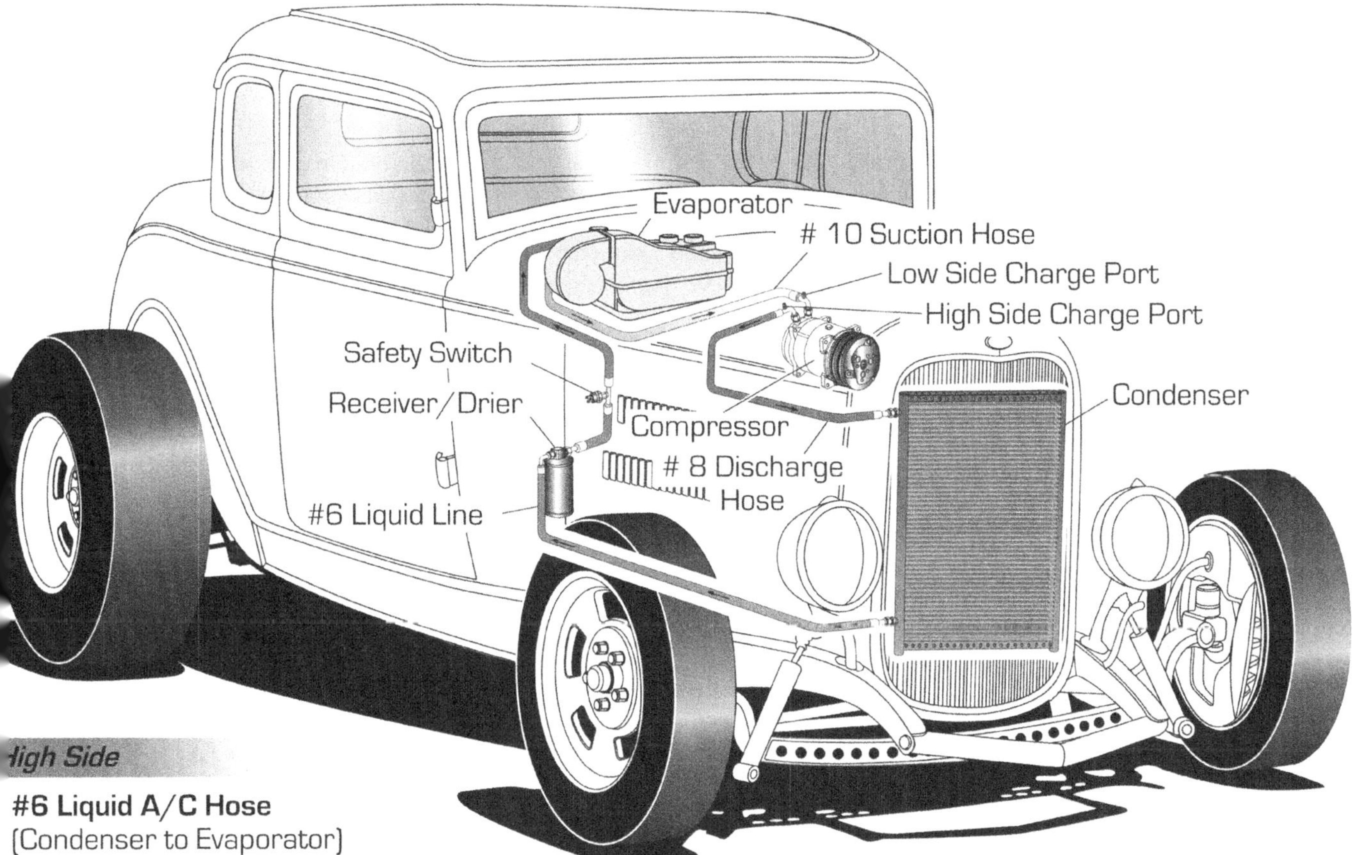

High Side

#6 Liquid A/C Hose
(Condenser to Evaporator)
#8 Discharge A/C Hose
(Compressor to Condenser)

Low Side

#10 Suction A/C Hose
(Evaporator to Compressor)

Direction of Refrigerant Flow Indicated By Arrows

CHAPTER TWELVE

Seal the Deal

Windows and Weatherstripping

By Ron Ceridono, Gerry Burger, and Tim Bernsau | Photography by Brian Brennan and Tim Bernsau

Since our bright red '59 Chevrolet rolled out of the paint booth at Hot Rods by Dean, the body is now buffed and ready for rubber. Unlike early cars of the '30s where weatherstripping was almost an afterthought, by the late '50s Chevrolet had actually designed weatherstripping into the body design of the car. Channels to hold the rubber were stamped in the body panels and formed rubber gaskets were designed specifically for each vehicle. A lot of engineering goes into designing a rubber profile to keep the air out of a vehicle

Our 2014 AMSOIL/*STREET RODDER* Road Tour '59 Chevy is taking shape. Dean Livermore and Wesley Zeller, of Hot Rods by Dean, are preparing to install the glass.

and also channel water properly away from doors, windshields, and decklids. That's the good news; the bad news is all rubber has a limited lifespan and now some 55 years later what was once supple and effective is now brittle, broken, and ripped.

Happily SoffSeal manufactures virtually every piece of rubber for our '59 Impala, from the front glass to the trunk seal. They use factory-style attachment methods for the formed rubber and when installed properly you have a nice, tight vehicle. And of course the look of brand-new rubber around door and decklid openings goes a long way to giving your car that detailed look.

1

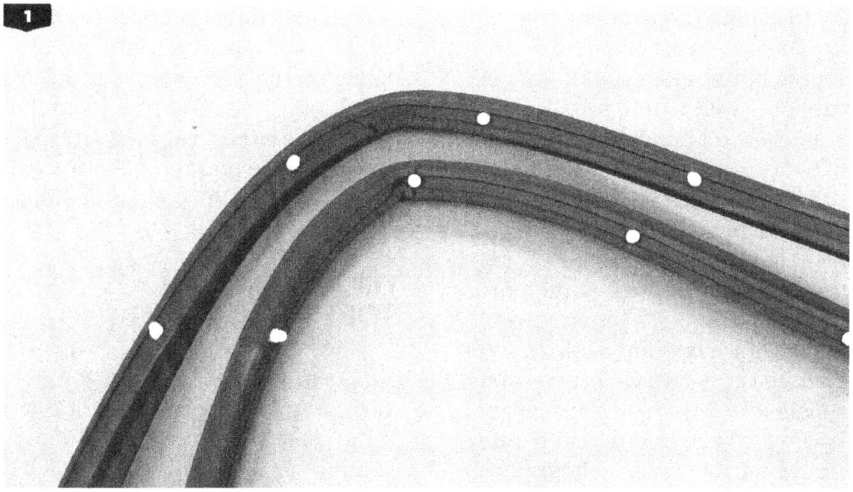

Every piece of rubber needed for our '59 Impala was sourced through SoffSeal. Note the factory-style mounting pins formed into the door seal. This ensures a perfect installation.

2

Starting with a nice, freshly painted car is the perfect way to begin installing weatherstripping. Sean Rosic began with the installation of the door rubber. By working carefully around the bottom of the door and then up to the top each plastic pin pushed into the factory drilled holes. There is no need for any rubber cement between the pins.

3

The door rubber is completely installed, avoid stretching the rubber between the pins, they should line up with just a slight bit of tension for a perfect fit.

Installing the weatherstripping is like many tasks on old cars, simple but not easy. Installing the rubber is well within reach of any home hot rod builder but it takes patience and some knowledge to do the job properly. We followed along with the professional staff at Hot Rods by Dean to clearly illustrate the proper installation of the pre-formed SoffSeal rubber products. Since we are working with a freshly painted car we have what we would consider to be a "new" installation. If you are replacing the rubber on an old car there will be hours spent cleaning off the old rubber residue before you can successfully install the new rubber. As always it is the initial preparation that makes the difference in the final product, so spending time on prep will save you time later and ultimately make for a perfect seal.

TIPS FROM THE SOFFSEAL PROS

Now, Hot Rods by Dean are professionals in every sense of the word, so they know all the tricks of the trade, but we thought it would be interesting to contact the professionals at SoffSeal to get some tips for the average guy building a car in his garage. Installing weatherstripping is a job most street rodders can handle and these tips will help to ensure a trouble-free installation. If you should need more advice SoffSeal tech support is available Monday through Friday to help you through any issues or concerns with their products, (800) 426-0902. We'd like to thank SoffSeal's Jeff Hubert for taking time to provide these helpful tips that range from purchasing weatherstripping to installing the product.

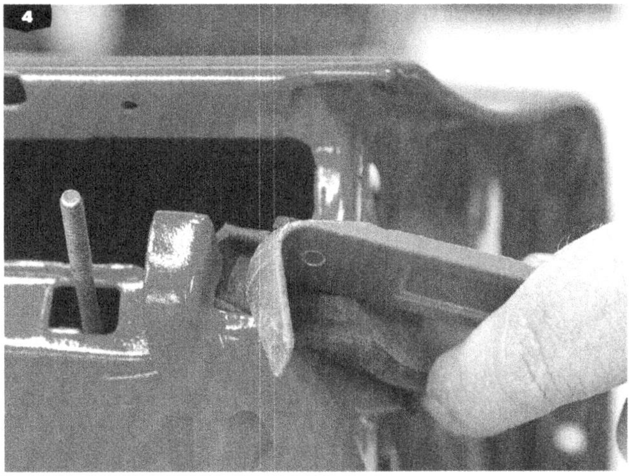

A plastic glass tool serves to apply pressure to the top pin as this one is a bit larger than the others and may require a little more pressure.

When the top pin snaps into the factory hole the SoffSeal rubber is securely anchored to the door. This piece was a complete snap-in installation with no cutting or gluing required.

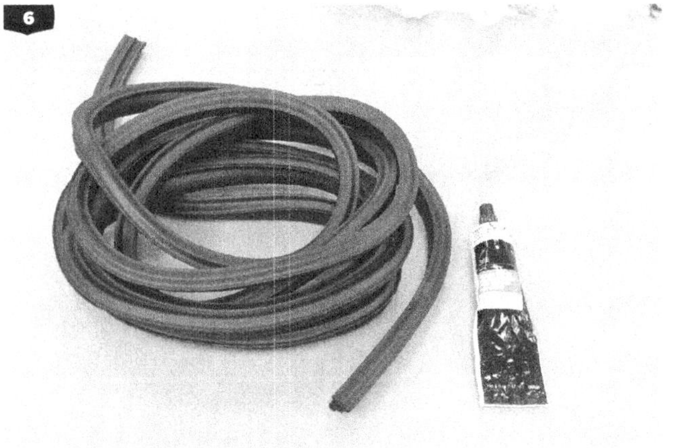

Next up was the trunk seal. This piece of weatherstripping has the exact factory profile so once again it fits perfectly to the contours of the decklid opening.

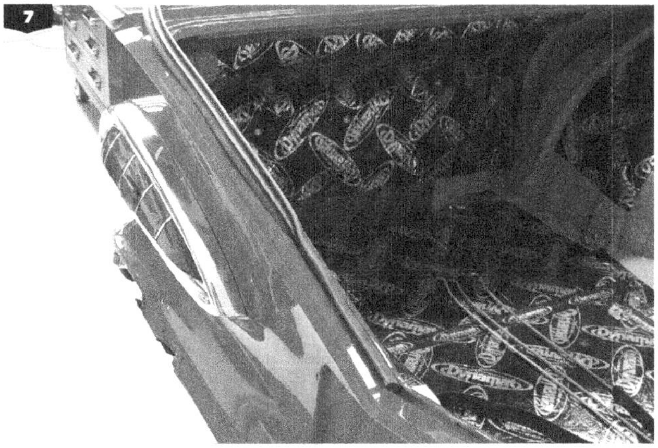

Placing the seam at the bottom of the "V" in the decklid ensures water cannot enter the trunk and follows the natural path of water. The weatherstripping is pushed into the channel in the body. A little soapy water will lubricate the rubber and make it slide into place easier.

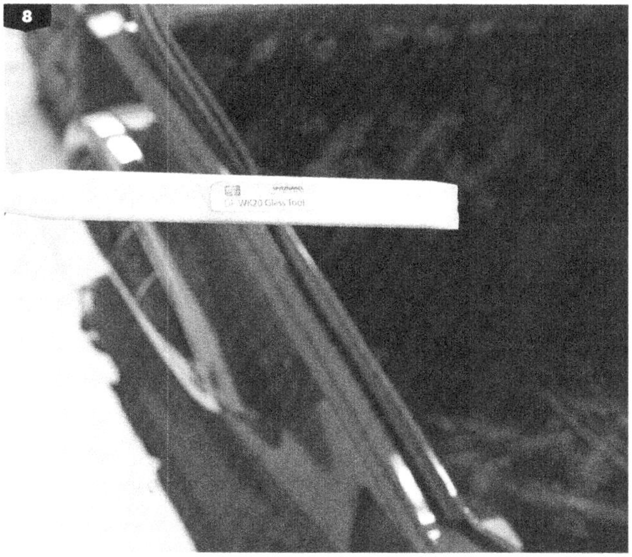

Having the right tool always helps. These nylon weatherstripping tools are available from most body shop supply stores. This particular tool is from Sptiznagel, DF-WK20 Glass Tool.

Using the glass tool the SoffSeal weatherstripping is carefully pressed into the channel. Work slowly and be careful not to push too hard as you can tear the rubber.

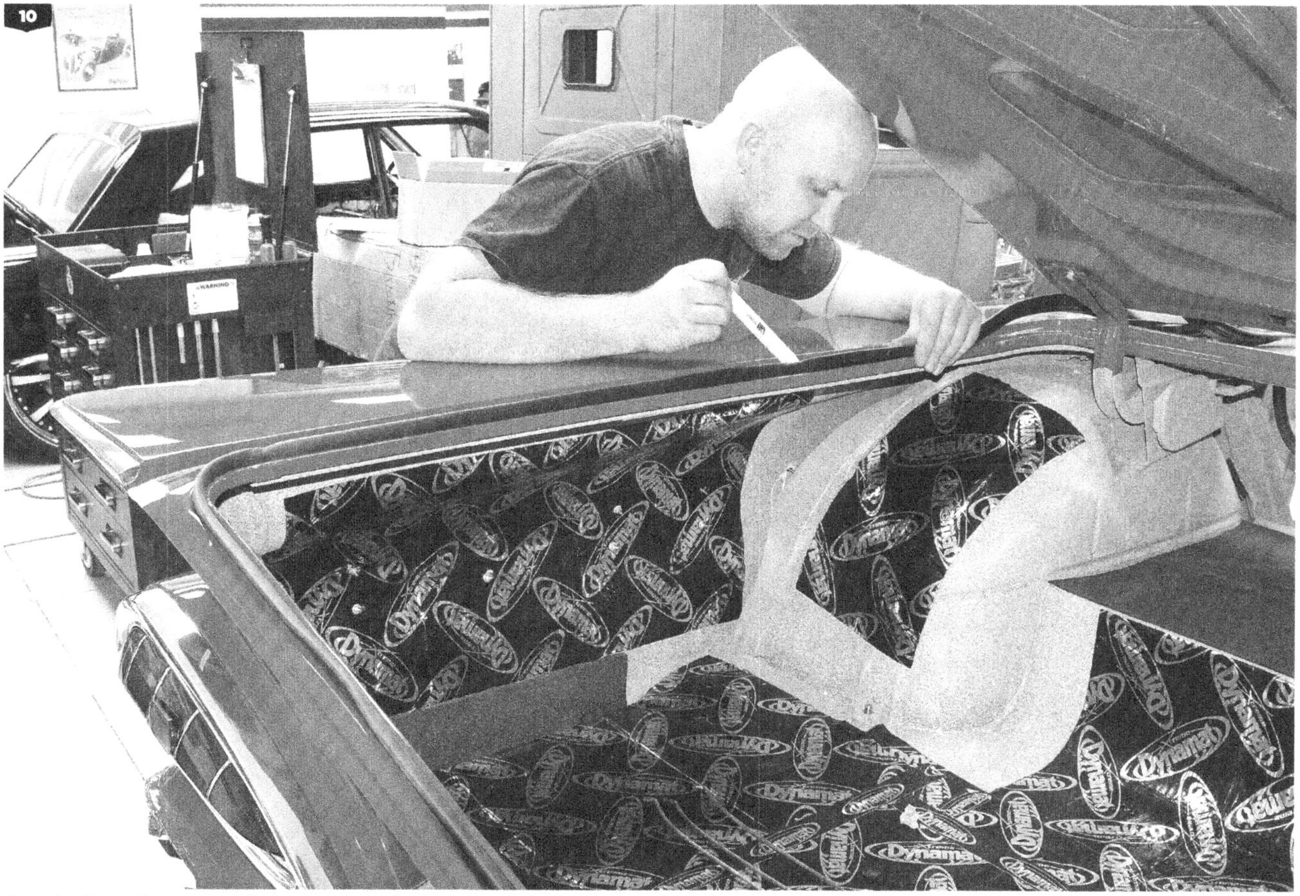

Sean Rosic continues to work the weatherstripping around the perimeter of the trunk. Slight tension is used to fit the rubber in the channel and the tool opens the profile so the top of the gasket is not inside the channel. Be careful not to stretch the rubber too much as it will lower the profile.

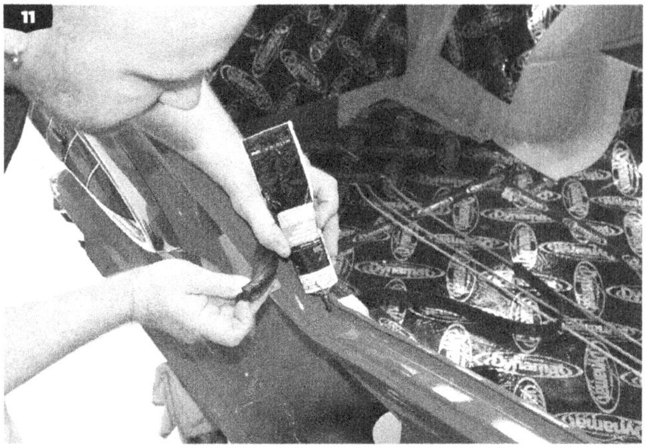

A small dab of 3M black weatherstripping cement is used at the seam and to anchor the ends of the weatherstripping. This will both cement and seal the only joint in the trunk weatherstripping.

With the doors and decklid sealed Zeller begins fitting the SoffSeal windshield rubber onto our Classic Industries tinted windshield.

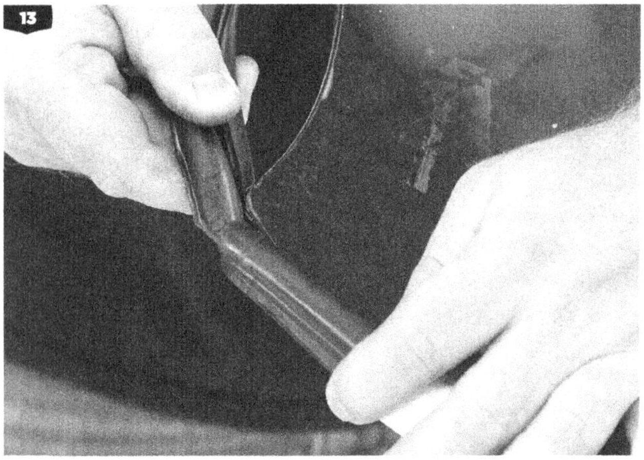

The windshield rubber is fitted to the glass dry, with special care taken to fit the formed corners to the corners of the windshield. Be certain the glass is all the way into the channel on the rubber.

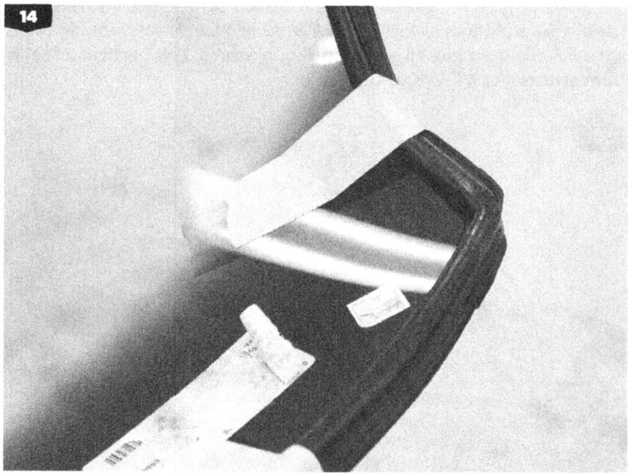

Pieces of masking tape hold the windshield rubber in place as the rest of the rubber is fitted to the glass.

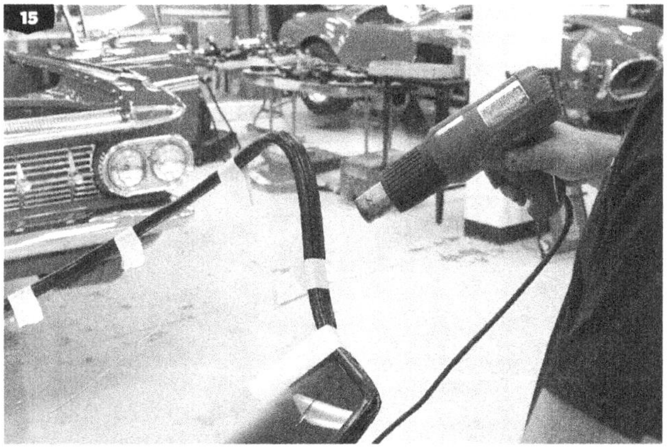

Where the rubber came around the lower dogleg portion of the glass a little heat was applied to help form the rubber to the contour. Be extremely careful using a heat gun on rubber weatherstripping as it is easy to melt or distort the profile.

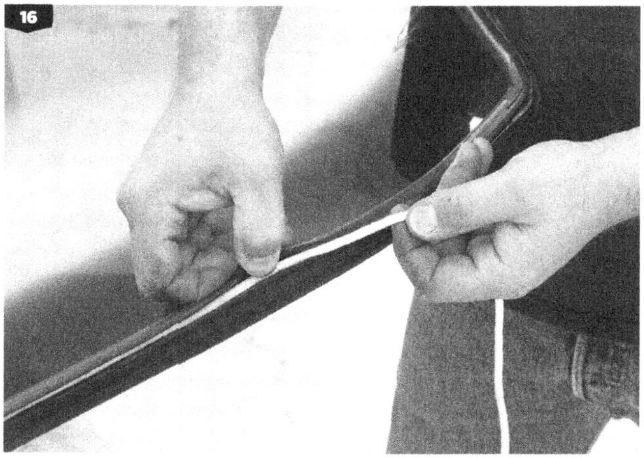

After the entire piece of rubber is fitted to the windshield, a piece of 1/4-inch rope is carefully inserted into the opening in the rubber. This opening fits over the inside lip of the windshield channel on the body.

The rope ends in the middle of the lower portion of the windshield. Masking tape holds the excess rope out of the way for now.

Livermore and Zeller position the windshield carefully in the opening, taking time to center the glass both horizontally and vertically in the opening.

After pushing the glass and rubber into place it is time to fold the inner lip of the rubber over the body channel lip. A spray of soapy water on the inside of the windshield serves as a lubricant for the rubber lip.

20

While Livermore is applying pressure to the outside of the glass Zeller is carefully pulling the 1/4-inch rope up and out, which curls the inside rubber lip over the channel. This takes a little practice and "feel" to get the rubber started.

21

By working together with the proper outside pressure and careful rope work on the inside, our Classic Industries windshield is held in perfect position by the SoffSeal rubber. Working the dogleg on a '59 Chevy windshield is a very challenging area. No adhesive or cement is used on the windshield rubber.

22

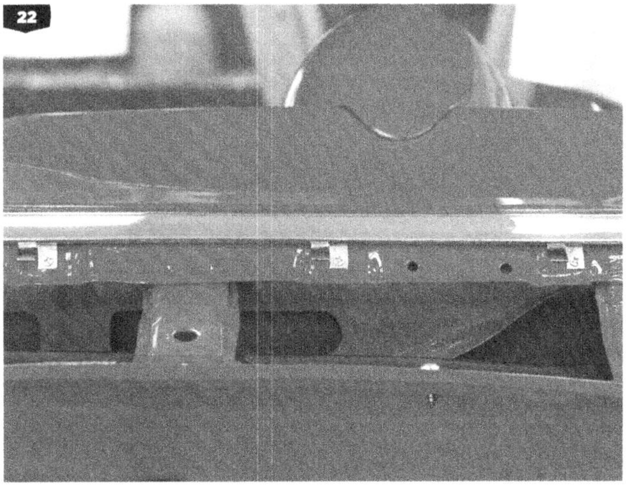

Next the stainless trim is pushed into the outside opening in the windshield rubber. This holds the molding in place and the molding expands the rubber to ensure a good seal. The stainless molding clips were sourced through Classic Industries.

23

Sherm's Custom Plating polished all the stainless trim for us so it looks better than new. Working the stainless piece in the corner of the windshield takes some pressure and patience, no hammers or excessive pressure, as this trim will dent easily.

24

With the windshield complete, attention was turned to the rear window on our '59 Chevrolet. We could have reused the old window but nothing looks better than new glass in a car, so we contacted Classic Industries for the rear window too.

25

After fitting the SoffSeal rubber to the rear window it was carefully centered left to right and top to bottom in the rear window opening. Once again no cement or adhesive is required for this installation.

26

The same rope in the inner slot of the gasket is used to slip the inner rubber lip over the body channel. This time the rope was ended in the top of the glass as that was the easiest place to begin the rope pull.

27

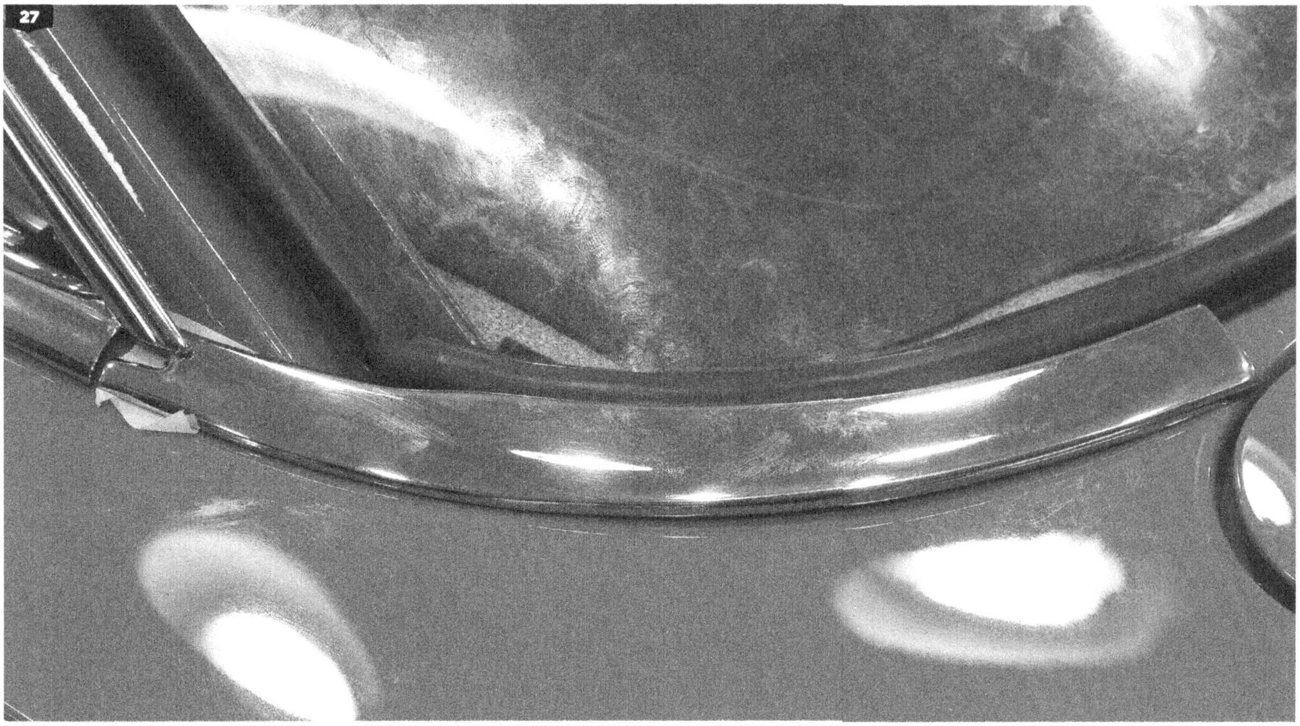

Once again the stainless trim is pushed into the rubber and fastened to the body with all new mounting hardware.

28

Wes from Hot Rods by Dean finished installing the Classic Industries garnish molding on the AMSOIL/*STREET RODDER* Road Tour '59 Chevy.

29

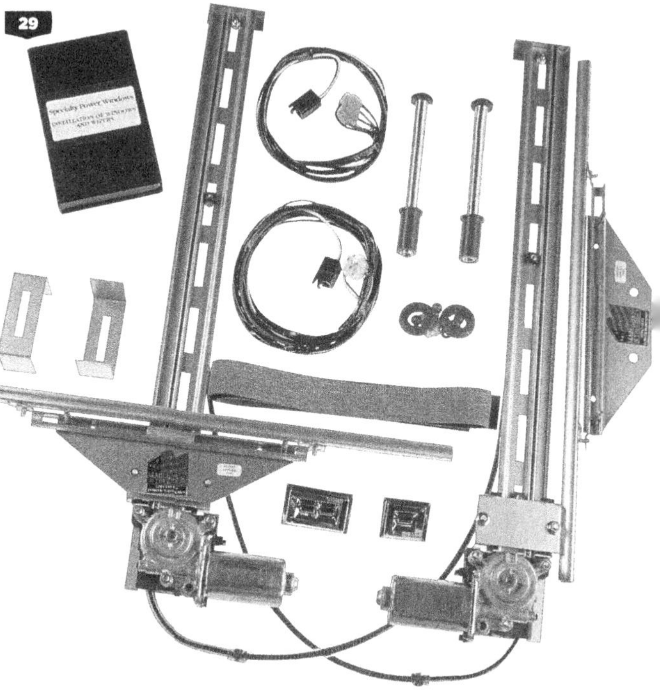

Electric power windows would provide Road Tour leader Jerry Dixey the luxury of lowering and raising the door windows with the touch of a switch. We used Universal kit PWL-2C from Specialty Power Windows. The system uses a GM-style motor and a vertical track designed to minimize window tilting, and is easily modified for a custom fit.

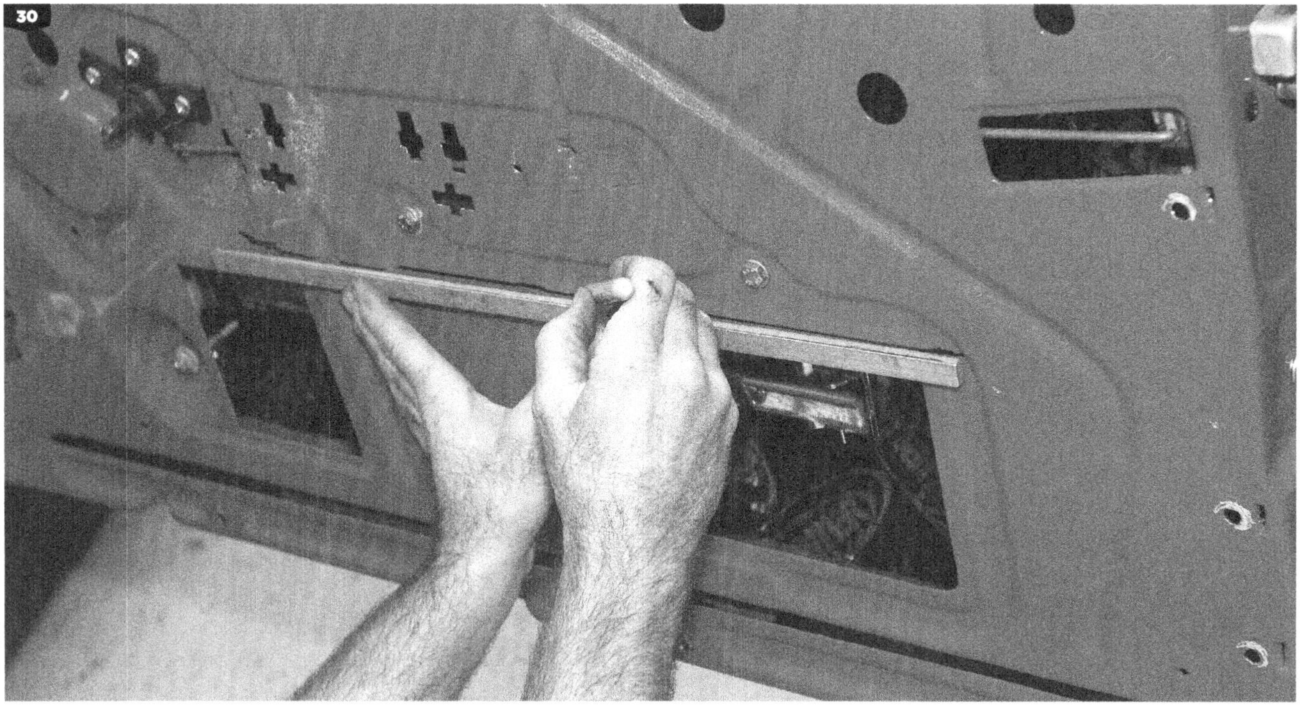

30

With the window glass in the full down position, Dru from HRBD marked the location of the bottom of the glass onto the door, and then removed the glass.

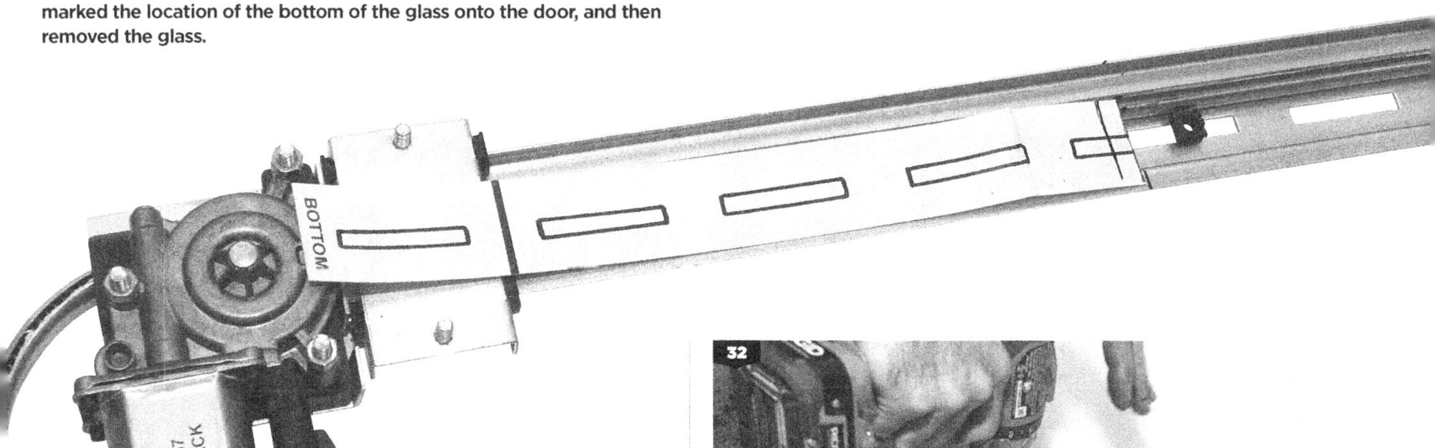

31

The paper template is cut, and used to determine the needed length of the regulator. This is how SPW ensures a custom fit for every application.

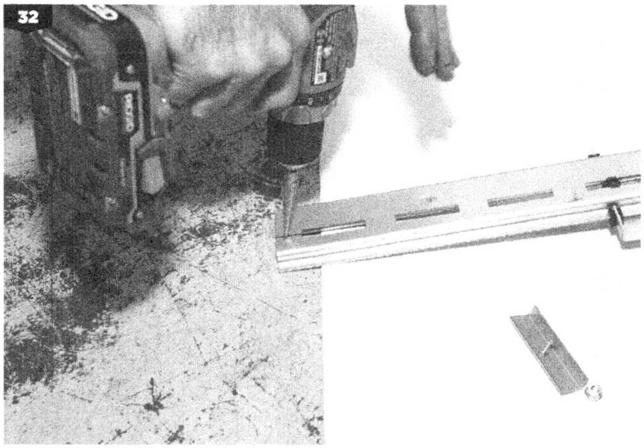

32

The endcap/upstop from the upper edge of the regulator is reinstalled at the "new" upper edge. If the regulator is cut across one of the vertical slots, the endcap bolt will fit in that slot. In this case, a bolt hole needed to be drilled.

33

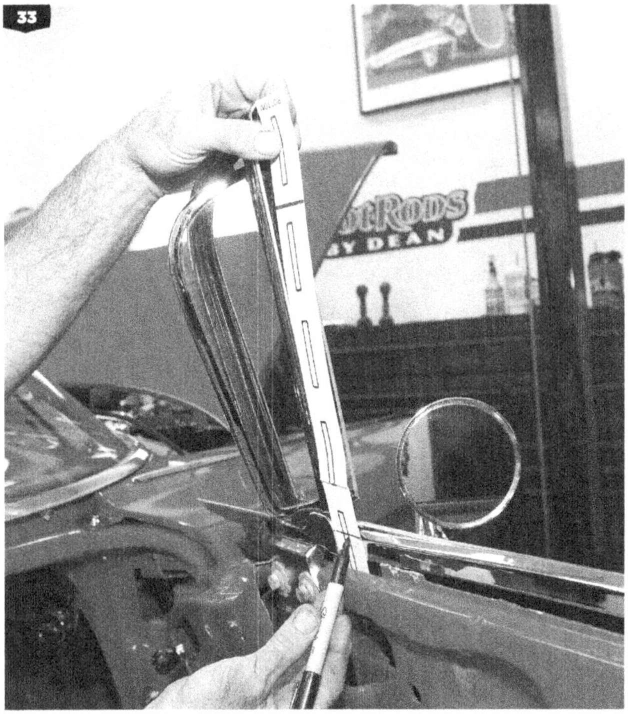

The Power Wonder Lift kit comes with a paper template, which is dropped into the door to the mark made in the previous step. The template is marked at the edge of the door, as shown, to measure the window glass travel from completely lowered to completely raised. Installation instructions call for adding 3/8 inch for adjustment.

34

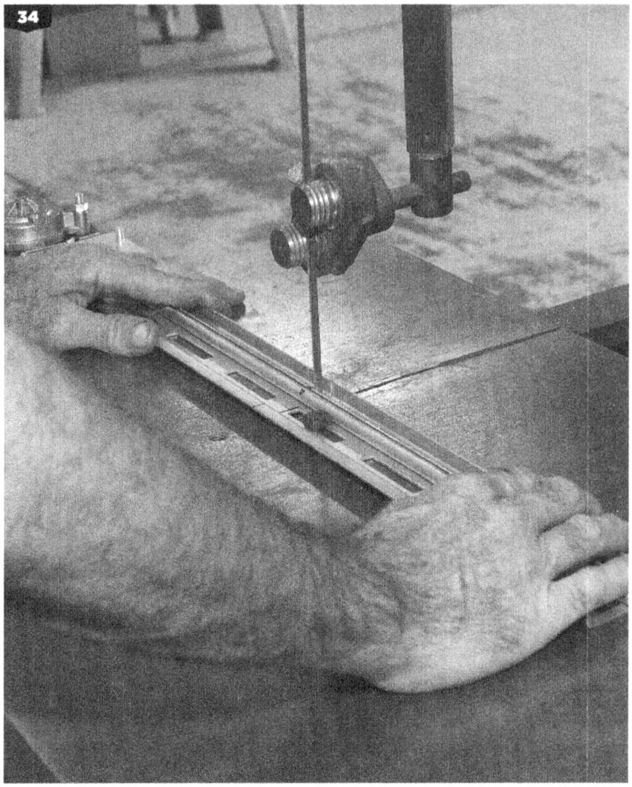

The vertical arm of the regulator is cut to the correct length as determined by the template.

35

The regulator should be positioned at the center of the window glass (the kit allows for adjustment). The door length was measured to figure out the correct spot.

36

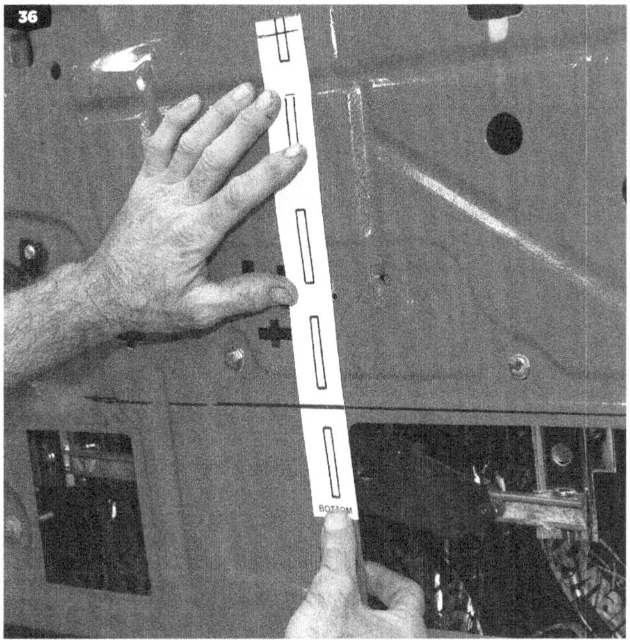

The regulator will be mounted on the inner doorskin. The cut paper template is used again to determine the best mounting points for the regulator along the center line. The regulator will be bolted to the door through the vertical slots.

Mounting holes are drilled in the door. The kit includes three-sided square flanged mounting brackets. These brackets can be welded or bolted to the inner door to provide stronger upper and lower mounting surfaces for the regulator. Horizontal slots in the brackets allow for more adjustment. Once in place, the bracket flanges can be removed.

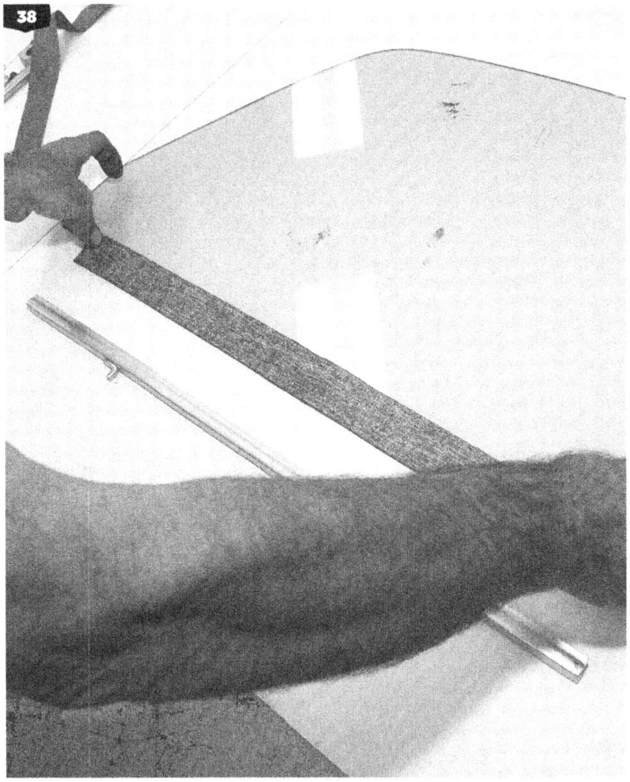

Glass-setting tape is provided with the kit. The tape is cut to the length of the lower glass channel and adhered to the glass.

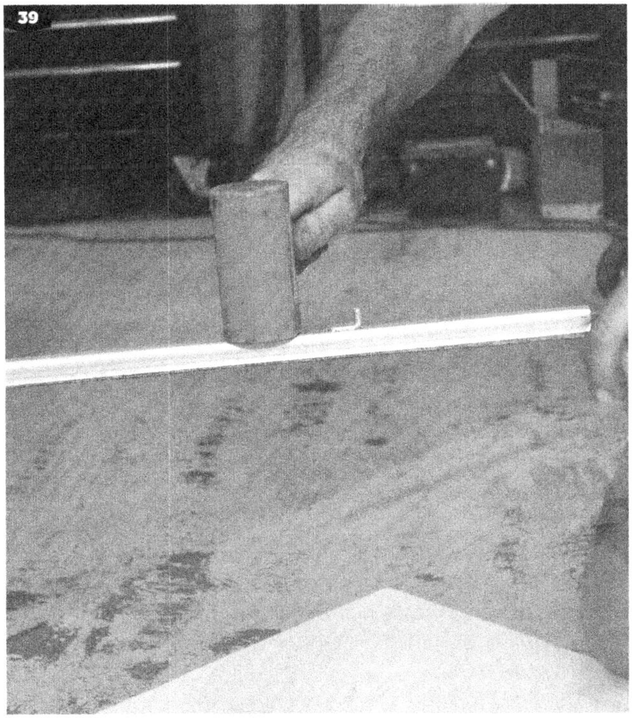

The lower channel is carefully fitted onto the window glass.

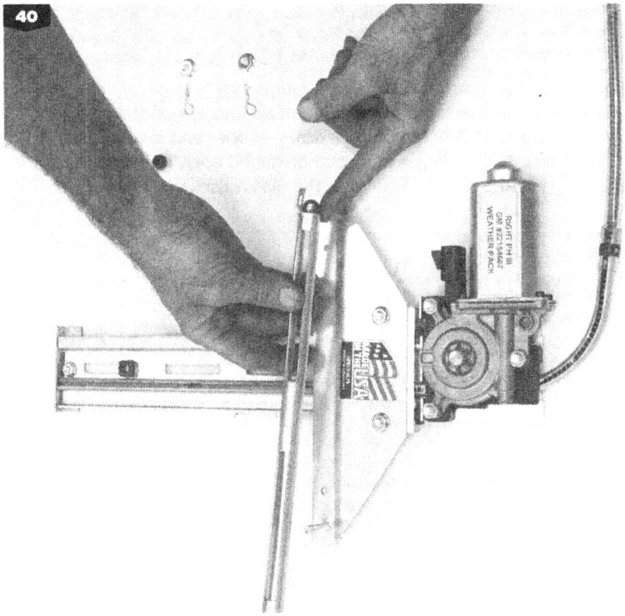

The regulator is assembled, using lithium grease on all rollers and slice channels. Dru is pointing to one of the nylon mounting balls that must be added to each end of the slotted tubes and fastened with cotter pins. The tubes are mounted on the lift bars. This design offers flexibility and prevents binding as the glass raises and lowers.

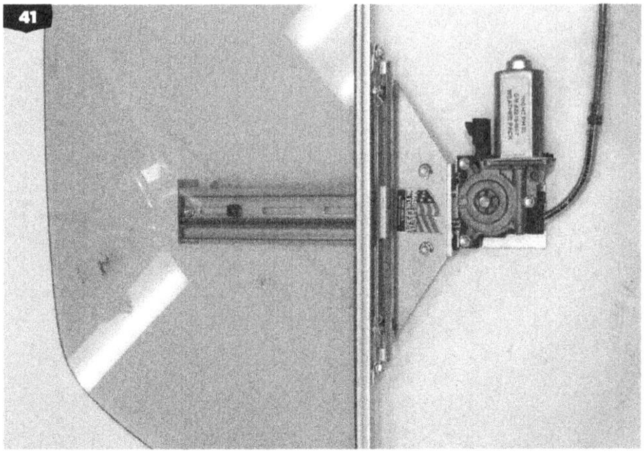

The lower channel is carefully fitted onto the window glass.

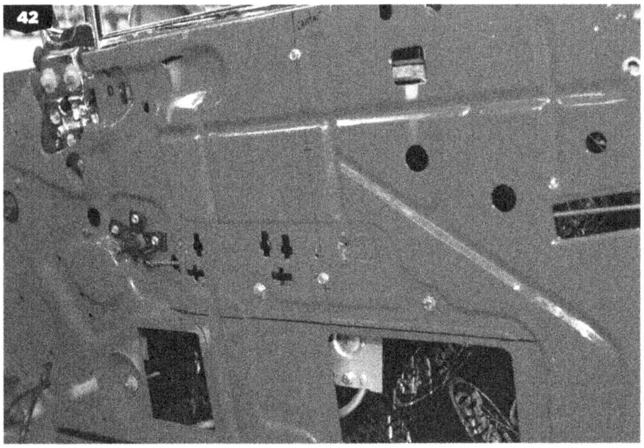

The regulator is assembled, using lithium grease on all rollers and slice channels. Dru is pointing to one of the nylon mounting balls that must be added to each end of the slotted tubes and fastened with cotter pins. The tubes are mounted on the lift bars. This design offers flexibility and prevents binding as the glass raises and lowers.

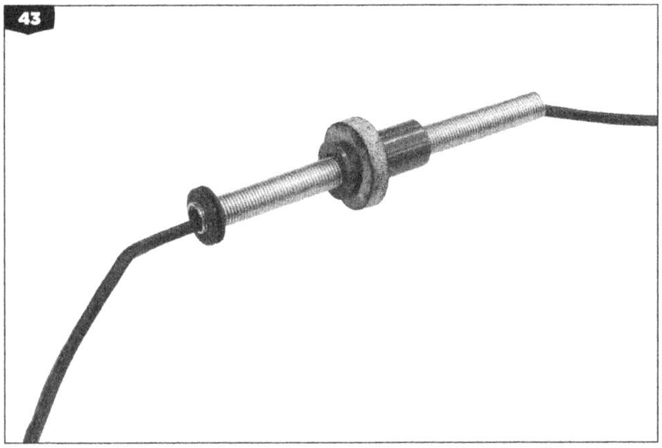

In addition to wiring harnesses and switches, and wiring diagrams, the kit comes with stainless steel flexible wiring conduits (and bushings and grommets) for running the wires through the doorjambs and hinge pillars.

Installing your new rubber

- Pegged door seals do not require the use of adhesive, though you can for a firmer, more solid fit.
- Weatherstrip adhesive should be applied like contact cement and not caulk.
- We recommend 3M 08008 Black Super Weatherstrip Adhesive Tube.
- Brush or spread a thin coat of adhesive onto both the rubber surface and the vehicle.
- Let the adhesive sit for about five minutes to get tacky. You will have plenty of working time to get the part on so be patient here.
- Fit the rubber onto the vehicle as it was test-fitted.
- If adhesive does seep out from the rubber, wipe it away immediately with a clean cloth and adhesive remover.

Installing universal rubber with adhesive backing

- Remove just 1-1/2 inches of the backing, starting from the same end you started from on your test-fit.
- Fold the backing to make a 90-degree angle.
- Place the rubber with the exposed adhesive onto the car.
- Using the exposed folded backing, slowly peel off more backing as you hold small sections of the rubber in place. Only peel about 2-3 inches at time.

Installing sliding rubber into a channel or onto an extrusion

- Spray the rubber and/or car with a little bit of lightly, soapy water. Just a bit of dishwasher detergent works fine (soapy not sudsy). This method is safest for your car and the rubber.

Prior to purchasing new rubber

- Know your car; is it a hardtop, sedan, coupe, or convertible? You would be surprised how many people are not sure of this.
- Has the car been modified? This will definitely effect fit of the rubber.
- Is the current rubber original? Always save the original rubber that came off of the car in case you need to check measurements and profiles.

Prior to installing new rubber on your car

- Test-fit the piece ensuring you have the correct piece (left/right, front/back).
- Also make sure that you are putting the seam in the correct place. Trunk seals will leak over time if you put the seam at the top.
- Trimming is sometimes necessary but usually not. Give us a call if you are not sure. We are happy to help!
- Prep the area where the rubber will go, especially if you are going to use adhesive. We recommend 3M Adhesive Remover 6041.
- There are a number of household solutions, but the 3M Adhesive Remover will not damage your paint.

Power Windows

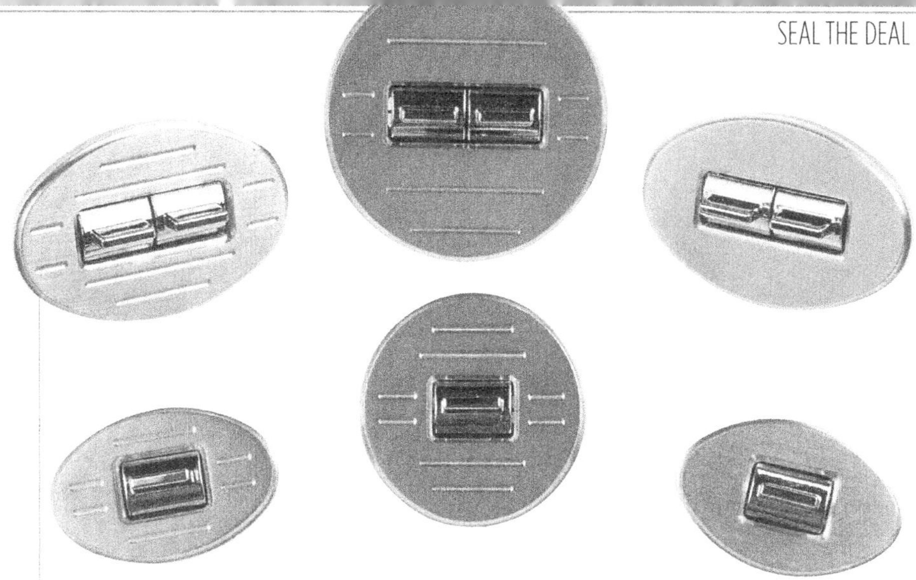

By the late '50s, fully electric power windows on were a factory option on practically every car rolling out of Detroit (on the '59 Chevy Impala, it was factory-installed option number 426). Even so, many of the cars being built as street rods, customs, and street cruisers probably came off the assembly line with old-fashioned, manually operated cranks windows.

Making the switch to electric power windows is one of the most popular and simplest upgrades, and the advantage of power windows appeals to even traditionally minded rodders. Whether you're raising it up to keep out the rain, or dropping it down to let in some fresh air or to ask a stranger for directions (just kidding), tapping a switch at your side is easier (and sometimes safer) than reaching for that crank handle.

The original owner of the AMSOIL/*STREET RODDER* Road Tour '59 Impala neglected to order option 426 when he bought the car. Fortunately, his mistake can be fixed with an aftermarket power window kit.

We were at Hot Rods by Dean to photograph the installation of a kit from Specialty Power Windows. The Georgia-based company has been around for almost 30 years. They manufacture a variety of complete and partial kits for many popular post-war GM and Ford car and truck applications, in addition to Ford Model As, Tall Ts, and a universal kit. SPW's complete two-window Power Wonder Lift kit (PN PWL-2C) was selected for the Impala. Also known at the Street Rod Kit, this cable-driven GM motor kit includes everything you need: regulators, lower glass channel, GM switches, switch and motor plugs, wire harness, hardware, installation templates, and a stainless steel wire conduit kit for the doorjamb openings, along with printed and DVD installation instructions (not to mention a helpful tech support crew back in Georgia, reachable by phone or email).

The printed instructions that come with the Universal kit include illustrations of an early style full-frame door as opposed to the Impala's hardtop door with no upper window frame. The installation procedure is generally the same. Although the PWL 2C kit is universal, it is designed to provide a custom fit for your vehicle. Modification to the doors is minimal—basically just a few mounting holes and slots in the case of our '59 Impala application.

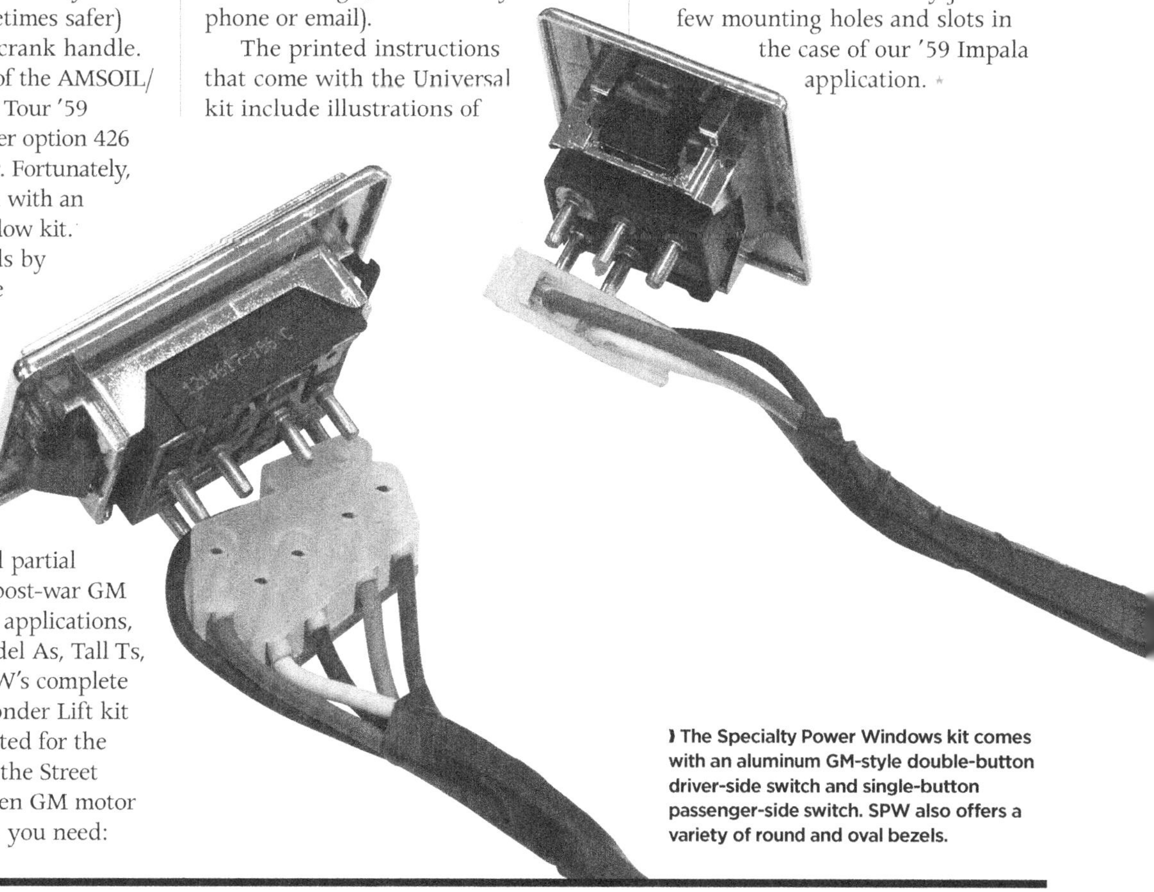

The Specialty Power Windows kit comes with an aluminum GM-style double-button driver-side switch and single-button passenger-side switch. SPW also offers a variety of round and oval bezels.

CHAPTER THIRTEEN

Interior Decorating

Upholstery

By Ron Ceridono | Photography by Brian Brennan

When our '59 Chevy first rolled through the door at Hot Rods by Dean (HRBD) to begin the transformation from raggedy to Road Tour ready it was missing the engine, trans, glass, some trim parts, and virtually everything inside the passenger compartment.

While the Impala now rests on a high-tech fully independent Roadster Shop chassis and has a full-on Ron Shaver engine backed by a Hughes overdrive automatic transmission, the exterior could best be described as being done in resto-rod style.

Designing the interior of the AMSOIL/*STREET RODDER* Road Tour Chevy was a unique challenge. Like the rest of the car it had to incorporate contemporary elements with the classic feel of 1959, an assignment that went to Glenn Kramer at Hot Rod Interiors by Glenn. Working at his craft since 1980, Kramer opened the shop that bears his name in 1983.

Building the interior for our

Chevy was an international effort of sorts. Maelle Riou is a talented young lady from France who has been working with Kramer in what could best be described as an apprenticeship program. The pair of stitchers didn't have much to work with, as the interior was gutted. Up front a pair of SCAT bucket seats were modified to fit while the center console and the trans tunnel were both modified in order to fit together. Modifications to the trans tunnel also included a recess to house the Lokar shifter. In the rear the seat springs were modified to fit the modified floorpan made necessary by the new perimeter frame.

To keep the passenger compartment comfortable and quiet, all the surfaces were covered with Dynamat and Dynaliner. In an effort to maintain the vintage look, the floor was covered in red factory-style loop pile carpeting while the seats and side panels were covered in a combination of fire red vinyl gray chevron pattern inserts. The headliner was made from red perforated vinyl.

Adding to the comfort level of the Impala is a tilt steering column from Flaming River and a complete stereo system from Custom Autosound. Classic Industries supplied all the interior trim, mirrors, and a long list of replacement parts. Thanks to Kramer and Riou's efforts the interior of the AMSOIL/*STREET RODDER* Road Tour Chevy was exactly what we were after. When sewing machines came to a halt and the last panel was put in place, 231 hours had been invested in rejuvenating the Impala's innards; we'd say it was time well spent. *

enn Kramer at Hot Rod Interiors by Glenn ed the Road Tour interior with a combination of age and contemporary elements. Topping the ning River tilt column is 15-inch Impala steering eel from Speedway Motors; instruments are ota Digital.

1

SCAT supplied the cool wrap-around bucket seats. The modified floor and the Chevy's low roofline created some challenges for Kramer.

2

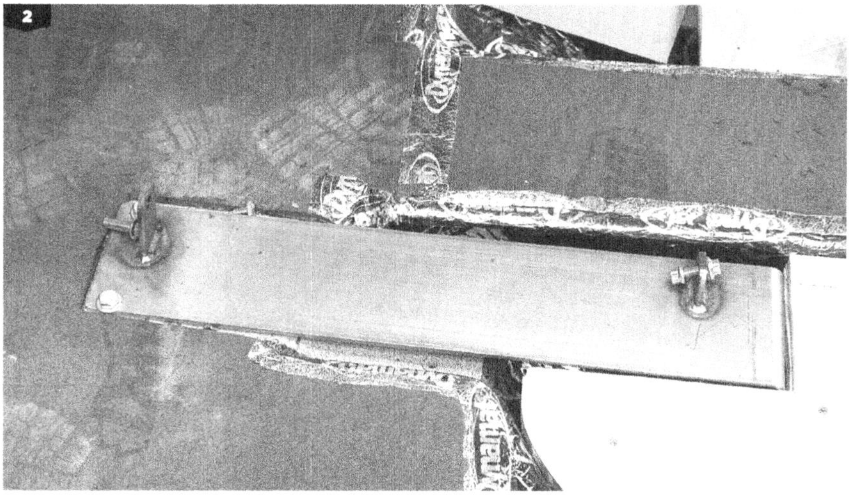

To lower the seats HRBD made custom mounting brackets. The SCAT seat adjuster framework attaches to the tabs.

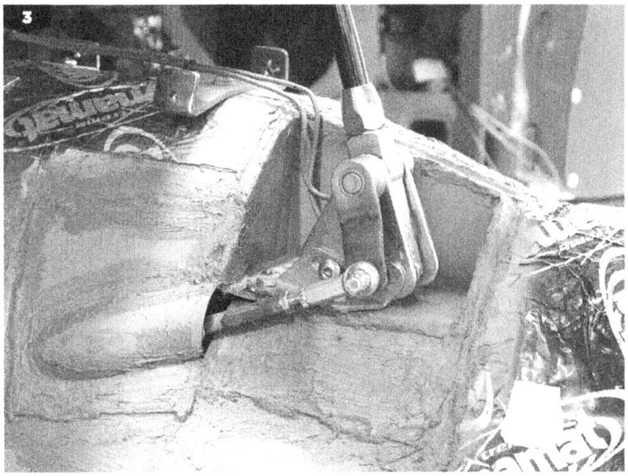

In order to get the Lokar shifter low enough to fit under the console that was planned, HRBD built a recess into the transmission tunnel.

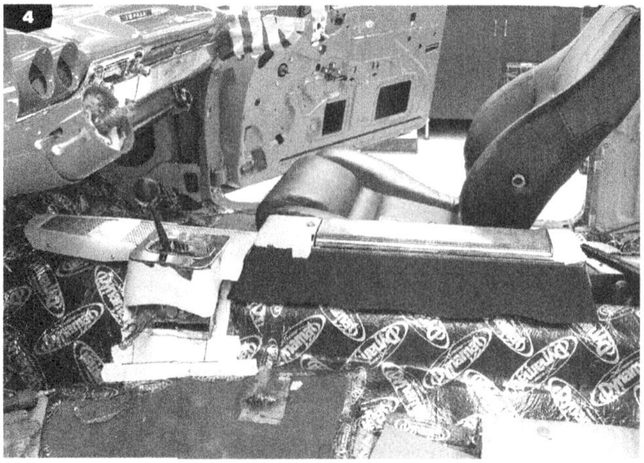

Between the seats is a reproduction '63 Impala console from Classis Industries. Careful trimming was required to join it to our modified floor.

One of the original styling elements Kramer wanted to retain was the relationship between the top of the seats and the doors. Consequently the decision was made to shorten the backrests.

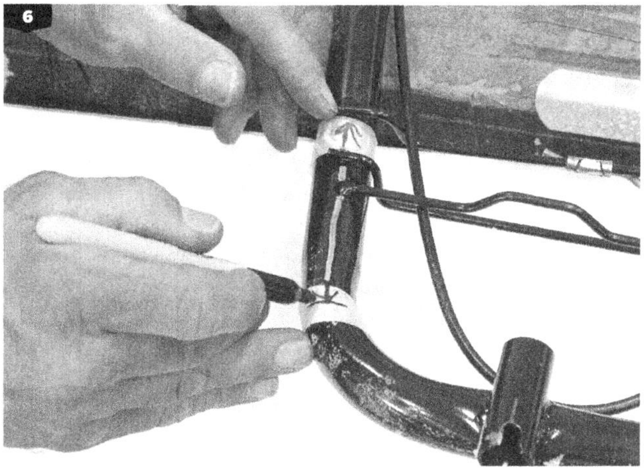

After removing the upholstery and foam, Kramer took a section our of the backrests' frames to shorten them.

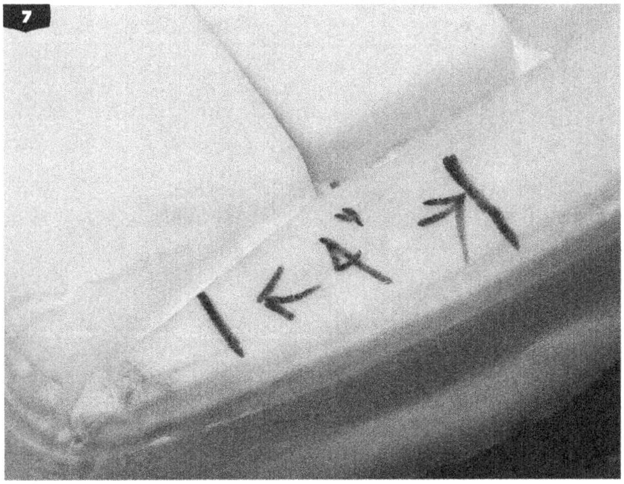

The same 4-inch section was removed from the foam.

Once the modifications were complete the seats were covered in a red vinyl and gray cloth. Gray was chosen to match the center of the wheels as well as the side trim.

The rear seat springs were from two-door post Chevrolet, which meant the bottom springs would have to be widened to fit the hardtop. They would also be notched to accommodate the modified floor.

Kramer had to shorten the seat back and widen it. After cutting the springs as needed they were spliced together with metal clips that were crimped in place.

Before the package tray was covered, a pair of 6x9-inch Custom Autosound speakers were put in place.

Before the headliner was installed the roof was treated to a layer of Dynaliner.

The new dome light, like all the other replacement interior trim pieces, came from Classic Industries.

Kramer spent many, many hours seated at his sewing machine to meet the completion deadline for the Road Tour debut. Despite the tight schedule the interior is flawless.

15

The new headliner was made from scratch by Kramer out of flame red perforated vinyl. The sun visors were covered to match (not yet installed).

16

Original-style loop pile carpet covers the floor, the Lokar shifter and modified center console are in place (note the Super Sport emblem). The pedal pads are from Lokar and Sherm's Custom Plating polished the aluminum trim.

17

Intern Maelle Riou gained valuable experience working on the Road Tour interior. Here she fits the custom foam cushions to the modified springs.

18

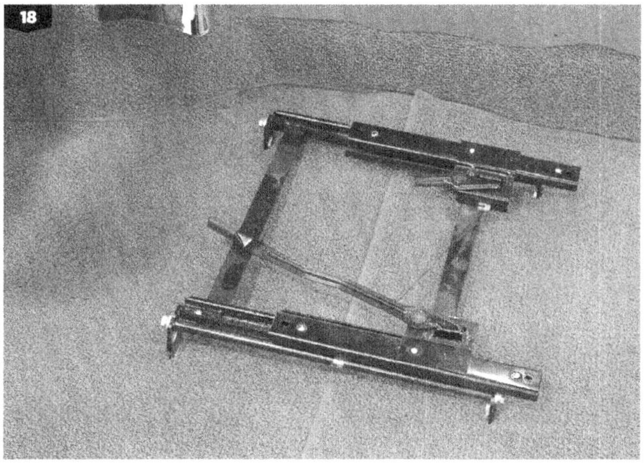

The SCAT seat adjuster assemblies bolt to the tabs on the brackets that were attached to the floor.

19

Bolted in place the modified seats are not only good looking but comfortable. Good thing as they will be occupied for 20,000-plus miles during the road tour.

20

Chrome trim adds some style to the door panels, the vinyl and cloth combination complements the seats.

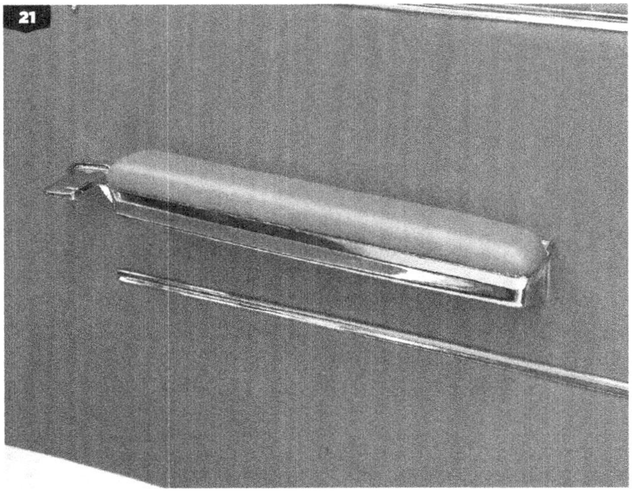

Those cool armrests and door handles are late '60s Chevelle items from Classic Industries.

In the rear, the two-tone and trim theme continues. Note the custom-built armrest.

Kramer and Riou did a masterful job of creating a stylish and comfortable rear seat with a semi-bucket seat look.

Included in the rear seat is the signature Impala speaker grille. Note the consistency of the surrounding stitching—that's quality workmanship.

Classic Industries supplied the reproduction Impala dashboard trim and the rearview mirror.

The cave-like trunk of the Chevy was done simply with the same loop pile carpet found in the passenger compartment. Nestled in the panel behind the rear seat is the Custom Autosound bass enclosure and amplifier.

CHAPTER FOURTEEN

Final Details

Compiled by Ron Ceridono | Information and Photography provided by the Manufacturers
Additional photography by Brian Brennan

From the introduction of the AMSOIL/*STREET RODDER* Road Tour 19 years ago the goal was to prove that a street rod built with aftermarket parts could be safe, reliable, and fun to drive—with our cars knocking off roughly 25,000 miles each year that point continues to be made. Certainly there can be no question that the continued growth in quantity and quality of aftermarket products, parts, and services keep our hobby alive and vibrant. Here are some of those that helped to make our 2014 Road Tour Chevy Impala a premium example of what a street rod can be.

AMSOIL

Lubrication for the Long Haul

There can be no better test of lubricants than we ask of those used in our Road Tour cars. They're on the road for long periods of time racking up 20,000 to 25,000 in a summer with few opportunities for maintenance and absolutely no time for breakdowns. These cars are always equipped with high-performance engines, driveline and chassis components that require the best in lubricants, that's why our Chevy uses AMSOIL synthetic products exclusively.

Synthetic oil has actually been around for quite some time, with the earliest attempts at production being made prior to 1900. The next big push in research was a result of Germany's oil shortage during World War II. It prompted intense investigation into the possibilities of creating synthetic lubricants. In fact, from 1944 on, 10 percent of the oil produced in that country was synthetic. However, it was the needs of jet engines that ushered in the new era of synthetics; long duration, high-speed operation, as well as huge variations in operating temperatures demanded performance beyond the scope of conventional oils. Synthetic oil was the answer to that need.

AMSOIL Z-Rod Synthetic Motor Oil meets API-SL and earlier specifications, allowing for the increased levels of anti-wear additives. It's recommended for a maximum drain interval of 5,000 miles or one year, or 3,000 miles/one year on heavily modified vehicles.

1) **AMSOIL Synthetic 10W-40 Motor Oil provides the extra anti-wear protection required by engines with flat-tappet cams and high-tension valve springs. 2) AMSOIL Synthetic Multi-Vehicle Automatic Transmission Fluid is engineered to outperform conventional automatic transmission fluids. It's recommended for transmissions, hydraulics, power steering, and other applications.**

AMSOIL Products

Synthetic 5W-30 SS Motor Oil

Signature Series Synthetic Motor Oil exceeds the higher performance demands of modern engines and withstands the stress of higher horsepower and higher heat applications. Signature Series Synthetic Motor Oil is engineered to outperform competitive conventional and synthetic motor oils. It delivers long-lasting performance and protection.

Absolute Efficiency Oil Filter

These filters feature advanced full-synthetic media that traps and holds a greater amount of small, wear-causing contaminants compared to conventional filters. They provide extended service intervals for increased convenience, while helping reduce engine wear with one of the best efficiency ratings in the automotive market, providing a filtering efficiency of 98.7 percent at 20 microns.

Synthetic Multi-Vehicle SS ATF

Signature Series Multi-Vehicle Synthetic ATF is engineered to outperform conventional automatic transmission fluids and provide outstanding performance and protection in the severe operating conditions.

Synthetic 75W-90 Gear Lube

Severe Gear Synthetic Extreme Pressure (EP) Gear Lube is specifically engineered for maximum performance in severe-duty applications. This state-of-the-art formulation maintains its viscosity for long-lasting protection against metal-to-metal contact.

Slip Lock Gear Oil Additive

This additive eliminates gear-housing chatter in cars, trucks and SUVs equipped with limited-slip, positraction and locking differentials. For use with both synthetic and petroleum gear lubricants and safely replaces manufacturer-recommended friction modifiers.

Antifreeze and Engine Coolant

Unlike conventional ethylene glycol-based products, which are acutely toxic, AMSOIL Antifreeze & Engine Coolant is formulated with propylene glycol. It is biodegradable and its low toxicity limits the threat to children, pets, and wildlife.

Dominator Coolant Boost

Formulated with proprietary tiered surfactant technology, providing

continued on page 130

3) AMSOIL Synthetic Water Resistant Grease meets the highest performance standards set by the National Lubrication Grease Institute (NLGI), GC for wheel bearing applications and LB for chassis applications. It may be used in automotive wheel bearings, chassis or other applications for which an NLGI No. 2 GC or LB grease is recommended.
4) AMSOIL Metal Protector (MP) and AMSOIL Heavy Duty Metal Protector (MPHD) are spray-on products that protect metal surfaces, displace water and silence squeaks. MP cuts through rust and corrosion to restore free movement of "frozen" hardware parts. MPHD lubricates metal surfaces, leaving a dry wax-like film.

AMSOIL Products

continued from page 129

quick and effective heat transfer inside radiator and cylinder heads, which results in reduced operating temperatures, more efficient operation, increased horsepower, and significantly reduced engine warm-up times in cold weather.

ORMD PI Performance Improver

This additive improves fuel mileage an average of 2.3 percent and up to 5.7 percent. Reduces emissions such as hydrocarbons (HC) up to 15 percent, carbon monoxide (CO) up to 26 percent and nitrous oxides (NOx) up to 17 percent. Restores power and performance. Reduces the need for costly higher-octane fuel.

Synthetic Multi Purpose Grease #2

This synthetic grease reduces friction and wear in high-speed, high-temperature components like wheel bearings (NLGI #2 is the primary wheel bearing recommendation). Keeps components running clean and trouble-free for outstanding performance and long life.

Synthetic Power Steering Fluid

AMSOIL Multi-Vehicle Synthetic Power Steering Fluid provides excellent wear protection. Its synthetic formulation delivers better lubricity and reduced friction, resulting in cooler operating temperatures, longer component life, and quieter operation.

Circle Racing Wheels

In 2005, Mike Stallings purchased Circle Racing Wheels and began moving the company forward into all facets of asphalt and dirt racing, which include winged Sprint Cars and Sprint Cup NASCAR racing. In addition to his goal of manufacturing and supplying the highest quality and the most competitive racing wheels available in the industry, Stallings also introduced a complete line of custom billet aluminum wheels for street rods and muscle cars—samples of same can be found on the 2014 AMSOIL/*STREET RODDER* Road Tour Chevrolet.

Our Impala rides on Circle Racing Wheels—up front they are 19 inches in diameter, 10 inches wide, with 5.25-inch backspacing, while in the rear the numbers are 20, 12, and 6 inches respectively. Tires are Michelin Pilot Super Sports, 285/35R19 fore and 345/30R20 aft.

❯ Circle Racing Wheels was responsible for the aluminum rollers found on all four corners of the '59. A car the size of our Impala wears large diameter wheels well.

Using their in-house equipment, Hot Rods by Dean took care of mounting and balancing the tires, not a particularly easy task when working with rubber this size.

Dean took a deep breath and hoisted a front wheel in place.

IDENTIFYING WHEEL DIMENSIONS

There are four important measurements that are used to categorize and identify a wheel for fitment purposes. They are:

A. RIM DIAMETER
This is the actual diameter of the wheel at the point where the tire bead seats (not the outer lip)

B. RIM WIDTH
Measure this from the inside of the outer lip at the bead seating point to the inner lip

C. BACK SPACING
A very important measurement, it's from the inside of the wheel at the point where it contacts the hub, brake drum or axle flange to the inside edge (lip) of the wheel

D. BOLT PATTERN
Count the number of mounting holes for the wheel and determine the diameter of a circle that would run through their center. On a 4, 6 or 8-lug wheel it's a direct measurement. On a 5-lug wheel, you can measure from the center of one hole to the OUTER edge of the hole diagonally across from it and get an approximate number that's very close to the bolt circle diameter.

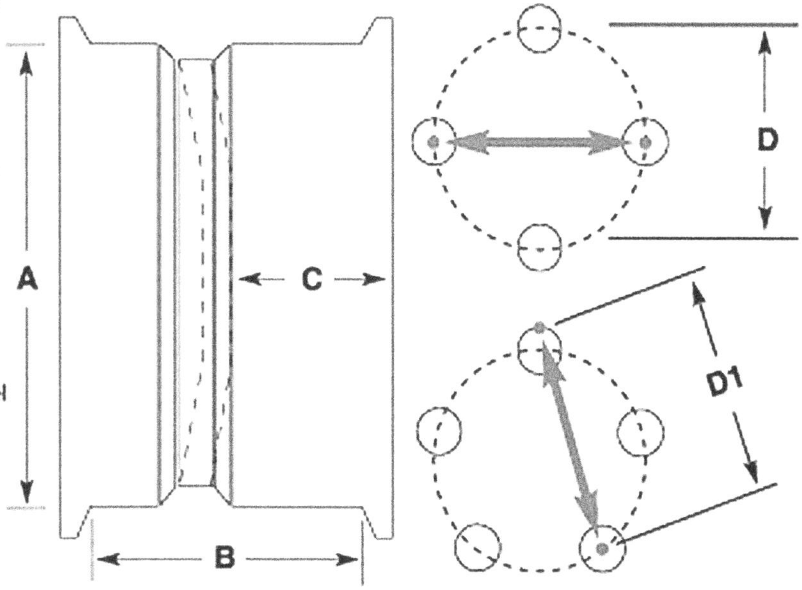

Measurements

Accurate measurements are essential to finding the right wheel for your application. Here's how we take the most common wheel measurements:

1. How do I measure backspacing?
Place the wheel (without tire) on a flat surface, face UP. Put a ruler down through the center hub hole. The distance from the flat surface to the bottom side of the center is the wheel backspacing.

2. What is offset?
Offset and backspacing are two terms commonly used in wheel measurement that can be confusing at times. Advertised wheel width is generally measured at the inside tire bead mounting surface, but you must determine the overall wheel width from the front to rear lip—usually 1 inch total. So a wheel that is advertised as 15Å-8 is 15 inches in diameter, 8 inches from tire bead to tire bead but 9 inches wide overall. Offset is the distance from the mounting pad to the centerline of the wheel. Offset can be positive (more backspacing) or negative (less backspacing).

Backspacing is the measurement from the mounting pad to the inner edge of the wheel. To order Circle Racing Wheels, you'll need to measure backspacing, not offset.

DETERMINING PROPER FITMENT

VEHICLE CLEARANCE

In many instances, the objective will be to fit a larger tire and within the confines of a given wheel well or fender. To make sure there is ample clearance for the tire and wheel (including considerations for suspension travel and brake drums/calipers), a number of measurements must be taken and supplied to the Billet Vintiques tech staff. Please refer to the drawing at right for references.

A ________ B ________
C ________ D ________
E ________ F ________

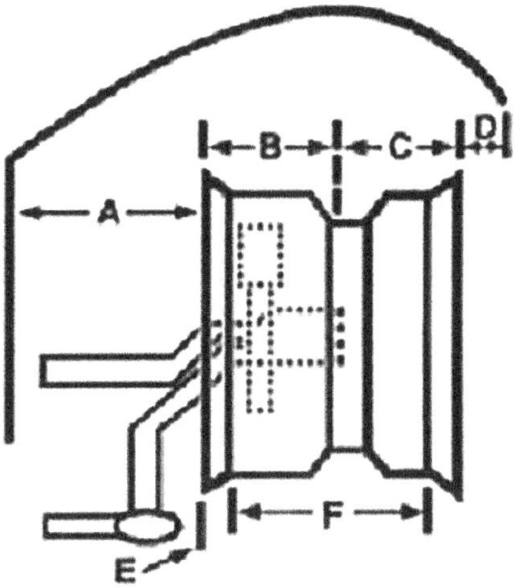

DISC BRAKES

The brand/model (aftermarket) or year/make/model of OEM discs, or the following dimensions will be required to provide necessary clearance for brake calipers:

A ________
B ________
C ________
D ________
E ________

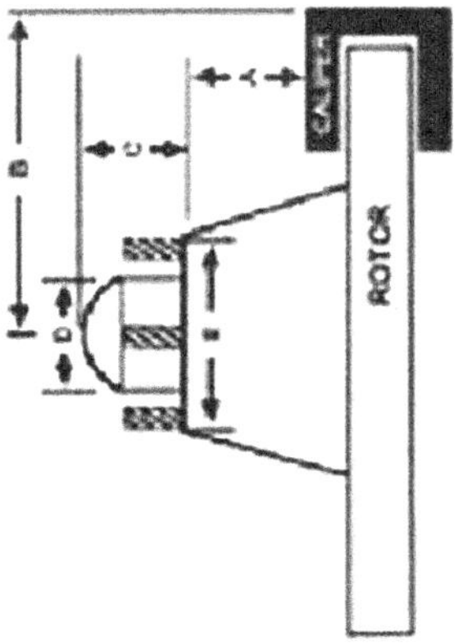

DRUM BRAKES

The year/make/model of OEM drum brake or following dimensions are required to assure proper fitment:

A ________
B ________
C ________
D ________
E ________

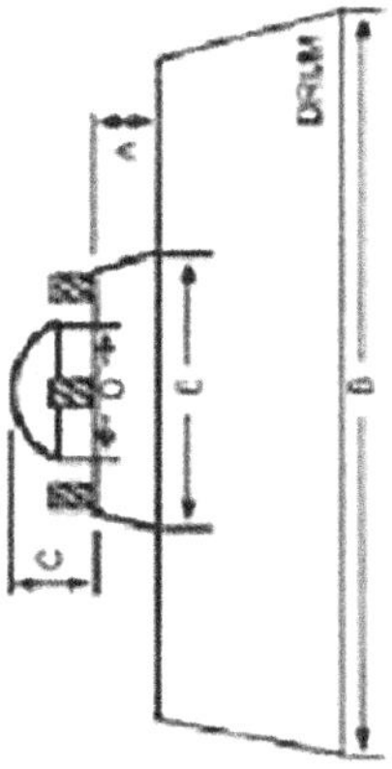

Dynamat

There are some products that make their presence known through sight and sound, and then there is Dynamat that goes unseen and certainly unheard, in the normal sense. To keep our Road Tour cars cool and quiet we routinely rely on products from Dynamat.

Dynaliner

AUTOMOTIVE

DESCRIPTION

Dynaliner is an ultra-lightweight closed cell high-performance insulation. Dark gray in color, Dynaliner is available in 1/8", 1/4" and 1/2" thicknesses. Dynaliner is self-adhesive with a high-temperature acrylic adhesive. Dynaliner is optimized for temperatures from -30°F to 200°F (-34°C to 94°C) and meets both UL and FMVSS flame resistance.

ACOUSTIC AND THERMAL PROPERTIES

Dynaliner is the perfect ultra-lightweight insulator to use on top Dynamat. This durable, crush and tear resistant material has the highest heat blocking properties available in a single layer synthetic foam-type material. Dynaliner is not affected by oil and does not absorb water.

APPLICATION

Easy to install, Dynaliner provides acoustic isolation and excellent thermal insulation for roof, interior firewall, floor, quarter panels, doors and even under hood. Dynaliner can be used in place of carpet pad.

INSTALLATION

Cutting: Dynaliner can be cut to a desired size and shape with a pair of scissors or razor knife. Dynaliner is self-adhesive. Make sure area is free from dirt oil and debris. For best results, work evenly from one side to the other applying even pressure.

AVAILABLE SIZES

Dimensions: 32" x 54" (81cm x 137cm)
Coverage: 12.0 ft² (1.1m²)
Available 1/8", 1/4" and 1/2" (3mm, 6mm and 6mm) thickness

TYPICAL MATERIAL PROPERTIES*

Part#	11101	11102	11103
Material Thickness:	0.125 in.(3.18mm)	0.250 in.(6.35mm)	0.500 in.(12.70mm)
Weight:	0.042lb./ft² (0.21kg/m²)	0.084lb./ft² (0.41kg/m²)	0.168lb./ft² (0.82kg/m²)
Density:	4.0lb./ft³ (64.6kg/m³)	4.0lb./ft³ (64.6kg/m³)	4.0lb./ft³ (64.6kg/m³)
Adhesive Strength:	tba	tba	tba
Tensile Strength:	tba	tba	tba
Tear Strength:	tba	tba	tba
Temperature Range: (Maximum)	-30°F to 200°F (-34°C to 93°C)	-30°F to 200°F (-34°C to 93°C)	-30°F to 200°F (-34°C to 93°C)
Temperature Range: (Optimal)	-30°F to 200°F (-34°C to 93°C)	-30°F to 200°F (-34°C to 93°C)	-30°F to 200°F (-34°C to 93°C)
STC:	N/A	N/A	N/A
FMVSS302:	Meets	Meets	Meets
UL Rating:	UL 94 HF-1	UL 94 HF-1	UL 94 HBF
R Value:	0.42°Fft²hr/Btu (0.07Km²hr/W)	0.83°Fft²hr/Btu (0.15Km²hr/W)	1.7°Fft²hr/Btu (0.3Km²hr/W)

* Material properties have tolerances of ±10% unless otherwise noted
Parentheses denote metric measurements.

The data provided in the material summary are typical of average values based on testing conducted by Dynamic Control or independent laboratories. They are indicative only of the results obtained in such tests and should be used for reference only. Materials used in situations not recommended must be tested under actual service to determine their suitability for that purpose.

DESCRIPTION

Dynamat Xtreme is a patented, light-weight, elastomeric, butyl and aluminum constrained-layer vibrational damper. Dynamat Xtreme conforms and fuses easily to sheet metal and other hard substrates. Material performance is optimized for temperature ranges between -10°C to +60°C (14°F to+140°F). Material can withstand temperature extremes between -54°C to +149°C (-65°F to +300°F) and is highly resistant to aging.

ACOUSTIC PROPERTIES

The acoustic loss factor "n" is used as a measure of ability to damp structure-borne sound. It states how much vibrational energy (in steel sheets for instance) is converted to heat rather than sound. For constructions containing several layers of damping material, the combined loss factor "n comb" is used. The theoretical maximum loss factor is 1 (no vibration). An undamped 1mm thick steel panel has a loss factor of roughly 0.001 at 200 Hz. Dynamat Xtreme applied to that panel would increase the loss factor to 0.417 @ +20°C (+68°F). Multiple layers of Dynamat Xtreme can be used to improve sound damping further.

APPLICATIONS

Dynamat Xtreme can be die cut to shape and placed onto the body surface after sheet metal cleaning operation and prior to paint system (typically at the sealer application operation) or on painted panels. Dynamat Xtreme is used as treatment for metal panels, partitions, ducts, doors, bins, panels etc. in railroad cars, buses, automobiles and ships. It is also used for ventilation ducts, relay cabinets, steel furniture, home appliances, sink units, computer equipment, machine tools and for many other purposes.

INSTALLATION

Dynamat Xtreme should be cut to the desired size before the release liner is removed. It may be cut with scissors, knife or die. Remove dust, grease, moisture, and other foreign matter from the application surface. Peel off the release liner. The simplest application technique is to bend the mat slightly and attach it along its shortest edge. The mat is then pressed firmly into place, preferably with a roller for larger pieces. This reduces the risk of leaving air pockets, which reduce the sound damping capacity. The temperature of the mat and application surface should not be below room temperature during fitting. Heating the material is not necessary.

SPECIFICATIONS

Appearance:
Black butyl based core with 4 mil aluminum constraining layer, craft paper release liner

Thickness:
0.067" (1.7mm)

Mass:
0.45lbs./ft.^{2} (2.20kg/m^2)

Acoustic Loss Factor @ Temperature (Using ASTM method E756 @ 200 Hz):

0.081 @ +14°F (-10°C)
0.240 @ +32°F (+0°C)
0.257 @ +50°F (+10°C)
0.417 @ +68°F (+20°C)
0.259 @ +86°F (+30°C)
0.194 @ +104°F (+40°C)
0.140 @ +122°F (+50°C)
0.094 @ +140°F (+60°C)

Temperature Range (Optimal Performance):
14°F to +140°F (-10°C to +60°C)

Temperature Range (Resistance):
-65°F to +300°F (-54°C to +149°C)

Adhesive Peel Strength:
42.6 lb./in. (74.8 N/cm) on cold steel

Chemical Resistance:
Resistant to water and mineral oils

Federal Standards Tests:
FMVSS 302: *Meets*

Handling And Application:
Material must be stored at room temperature for best application

Storage Information:
Number Of Sheets In Stack: 50 max
Material must be stored horizontally in its wrapping

DYNAMIC CONTROL
3042 Symmes Road • Hamilton, Ohio 45015
phone: 1-800-225-8133 • *fax:* 1-800-873-2423 • www.dynamat.com

Part No. 2204

REFERENCE CHART

4" x 10" XTREME DYNAMAT

Coverage: 0.28 ft.^{2} (0.03m^2)

Weight: 0.12 lbs. (0.05 kg)

Dimensions: 4"x10"
(100mm x 254mm)

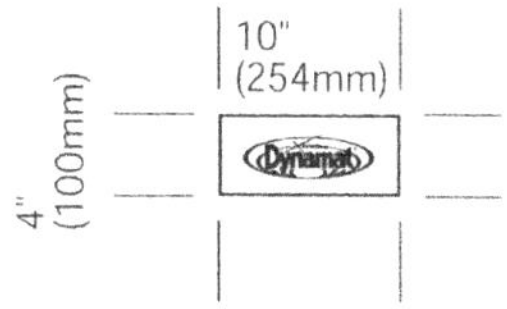

10" x 10" XTREME DYNAMAT

Coverage: 0.69 ft.^{2} (0.06m^2)

Weight: 0.31 lbs. (0.14 kg)

Dimensions: 10"x10"
(254mm x 254mm)

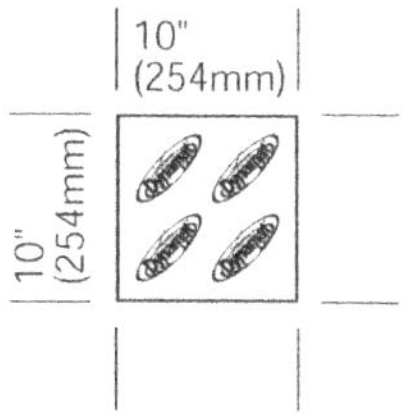

18" x 32" XTREME DYNAMAT

Coverage: 4.0 ft.^{2} (0.37m^2)

Weight: 1.80 lbs. (0.82 kg)

Dimensions: 18"x32"
(457mm x 812mm)

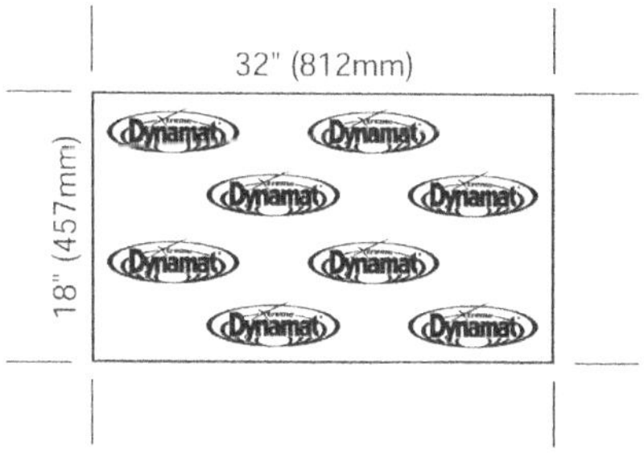

12" x 36" XTREME DYNAMAT

Coverage: 3.00 ft.^{2} (0.28m^2)

Weight: 1.35 lbs. (0.61 kg)

Dimensions: 12"x36"
(305mm x 914mm)

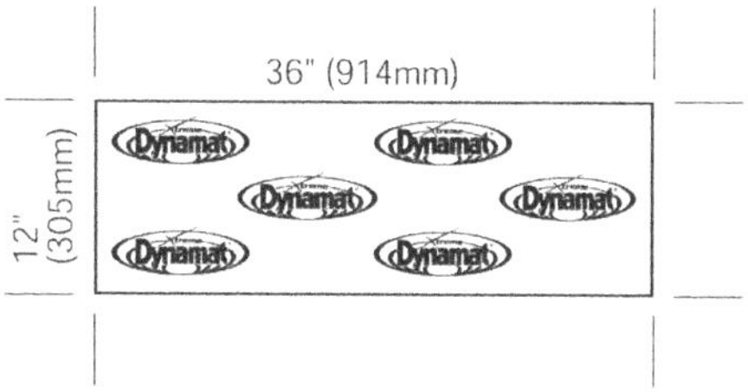

Note: Specifications subject to change without notice.

★14

Eastwood

Curt Strohacker worked in a service station as a kit to earn some spending, but he also repaired cars in his free time, buying and selling countless vehicles and learning valuable lessons about restoration. Strohacker also began to realize that the proper tools and techniques could save enthusiasts like him time and money. So, in the fall of 1978, Strohacker launched a business to better serve the burgeoning market of auto restoration and customization hobbyists, and he called it Eastwood.

The first Eastwood catalog was eight black-and-white pages of metal-finishing tools and equipment. Today Eastwood continues to publish a monthly, fullsize catalog along with full-service, ecommerce website, eastwood.com.

› Eastwood's combination shrinker and stretcher was an invaluable tool for making replacement sheetmetal panels and patches.

Flaming River

Like most cars of the era, our '59 Chevrolet came equipped with recirculating ball steering. As the name implies, this type of steering gear uses a series of ball bearings that roll between the gears in the steering box to reduce friction and wear.

While recirculating ball steering is still used, for most performance applications rack-and-pinion is the preferred design. A relatively simple design, a gear on the end of the steering shaft engages the teeth in a long bar called a rack and the ends of the rack are attached to the tie-rod ends. Turning the steering wheel makes the gear revolve, which moves the rack and ultimately the wheels. Rack-and-pinion steering is simple with few moving and pivoting parts, it's also precise and provides the driver with excellent feedback and road feel. But while it has obvious advantages, worm and roller and recirculating ball boxes do offer more mechanical advantage, so in most cases rack-and-pinion was used in lightweight cars

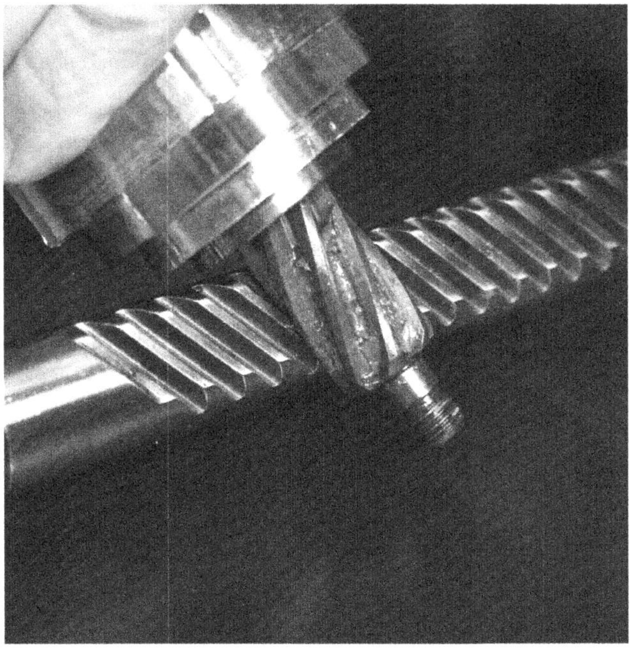

❱ The basics of a rack-and-pinion steering gear: The pinion gear is attached to the steering shaft, the rack converts the rotary motion of the steering wheel to the linear movement required to turn the wheels.

❱ The spool valve mounts to the rack—it controls the fluid from the pump that supplies the power assist to the rack. Like all the other components, Flaming River builds these in-house.

while a conventional gearbox was found in heavier vehicles. That was true until power racks came along. Power racks are now found in a number of larger cars, now including our '59 Chevy.

Choosing the Right Rack

Flaming River offers rack-and-pinion steering in manual and power designs for front and rear steer applications in a wide variety of configurations, so finding the proper gear for any application isn't a problem.

While there are exceptions, most aftermarket IFS applications that require a front steer rack use a Ford-style gear. There are two versions of this steering, the '74-78 Mustang II style with 16 inches between mounting points and the '79-93 Mustang style with 15-1/2 inches between the mounts.

It should be noted that in addition to the variations in mounting dimensions, steering gear travel is also a consideration when selecting a rack—too little travel and the car's turning circle will be larger than necessary, too much travel and there may be interference between the tires and the chassis on hard turns. While light rubbing between the tires and chassis may happen occasionally in extreme circumstances (parallel parking, turning in a driveway, and hitting a bump simultaneously) repeated heavy contact must be avoided as it can cause serious damage to components. In some cases when a long travel rack is used steering stops will be required.

Simple-and-Effective Rack-and-Pinion Steering

Our Ford was originally equipped with what is called worm and roller steering. While it was an improvement over the worm and sector style used earlier, eventually Ford, and the rest of the automotive industry, adopted recirculating ball steering. As the name implies, this type of steering gear uses a series of ball bearings that roll between the gears in the steering box to reduce friction and wear. While this design made steering easier, it also dramatically increased the service life of the steering gear as a result of the increased wear surface supplied by the balls rolling between the friction points.

The next big advancement in steering was the rack-and-pinion. A relatively simple design, a gear on the end of the steering shaft engages the teeth in a long bar called a rack and the ends of the rack are attached to the tie-rod ends. Turning the steering wheel makes the gear revolve, which moves the rack and ultimately the wheels. Rack-and-pinion steering is simple with few moving and pivoting parts, it's also precise and provides the driver with excellent feedback and road feel. But while it has obvious advantages, worm and roller and recirculating ball boxes do offer more mechanical advantage, so in most cases rack-and-pinion was used in lightweight cars while a conventional gearbox was found in heavier vehicles. That was true until power racks came along. Power racks are now found in a number of larger cars. Thanks to the lightweight engine, Honest Charley Garage opted for a manual rack for the AMSOIL/*STREET RODDER* Road Tour '51.

Installation Tips: Universal Joint Installation

1. Measure and mark 7/8 inch in from the end of the steering shaft
2. Position yoke onto shafting to the 7/8-inch mark. Tighten setscrews to mark shaft
3. Remove yoke
4. Using a 1/4-inch drill bit, countersink the marked position on the shaft
5. Apply red thread locker to threads
6. Reinstall universal joint onto shaft
7. Using a torque wrench, tighten the setscrews to no more than 25 ft-lb
8. Tighten jam nuts

Note: Inspect U-joints frequently to ensure tightness and proper fit

FAQs

Q: What different styles of columns does Flaming River offer?

A: Flaming River offers a variety of Tilt Columns and Non-Tilt Columns available in a wide selection of lengths and sizes. Floor Shift Tilt, Column Shift Tilt, and Tilt Key Columns are our most popular. Key columns can be made in either a Floor Shift or Column Shift style. Our columns come in three finishes: Paintable Mill Finish, Black Powdercoat, and Polished Stainless. Output shafts come in 1 inch to 48 splined shaft, 3/4 inch to 36 splined shaft, and 1 inch DD shaft.

Q: What is the best way to mount a steering wheel to a Flaming River column?

A: Flaming River has a wide assortment of Steering Wheel Adapters to fit original or aftermarket steering wheels to our columns. The universal five- and six-bolt wheel adapters contain the pattern for both five- and six-bolt steering wheels, so getting the perfect fit is even easier. Certain steering wheels, like the WaterFall, even have their own adapters to mount to a Flaming River column or original stock column. We also offer three-bolt adapters, nine-bolt adapters, and a number of adapters for Corvette and Mopar applications.

Flaming River Parts List	
PN FR20006	32-inch Paintable Tilt Column, 2-inch O.D.
PN FR2517DDPI	1DDx3/4 DD SS UJ POL
PN FR2516DDPL	34DDx3/4 DD SS UJ POL
PN FR20114	Factory Drop
PN FR20101	ASM MNT Floor Ring Split ø2.0
PN FR1850-22SSPL	22-inch 3/4-inch DD SS POL Shaft
PN FR1503	'79 Mustang Rack-and-Pinion
PN FR1600	Mounting Bushing
PN FR1601	'74-78 Mustang Outer End

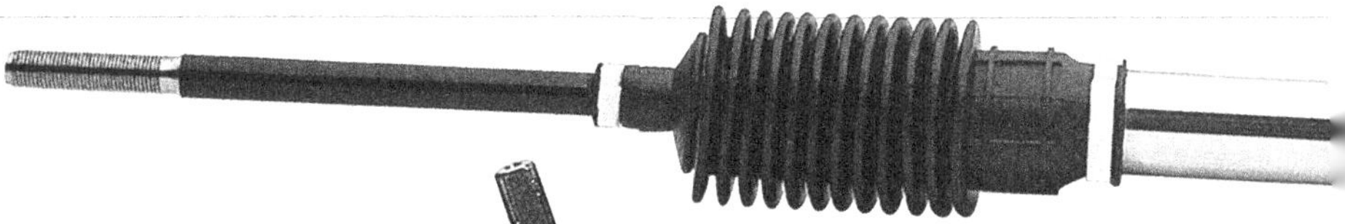

Flaming River builds rack-and-pinion units in a variety of configurations for both front and rear steer applications.

Building a righthand-drive rod? Flaming River can supply a rack-and-pinion gear for that.

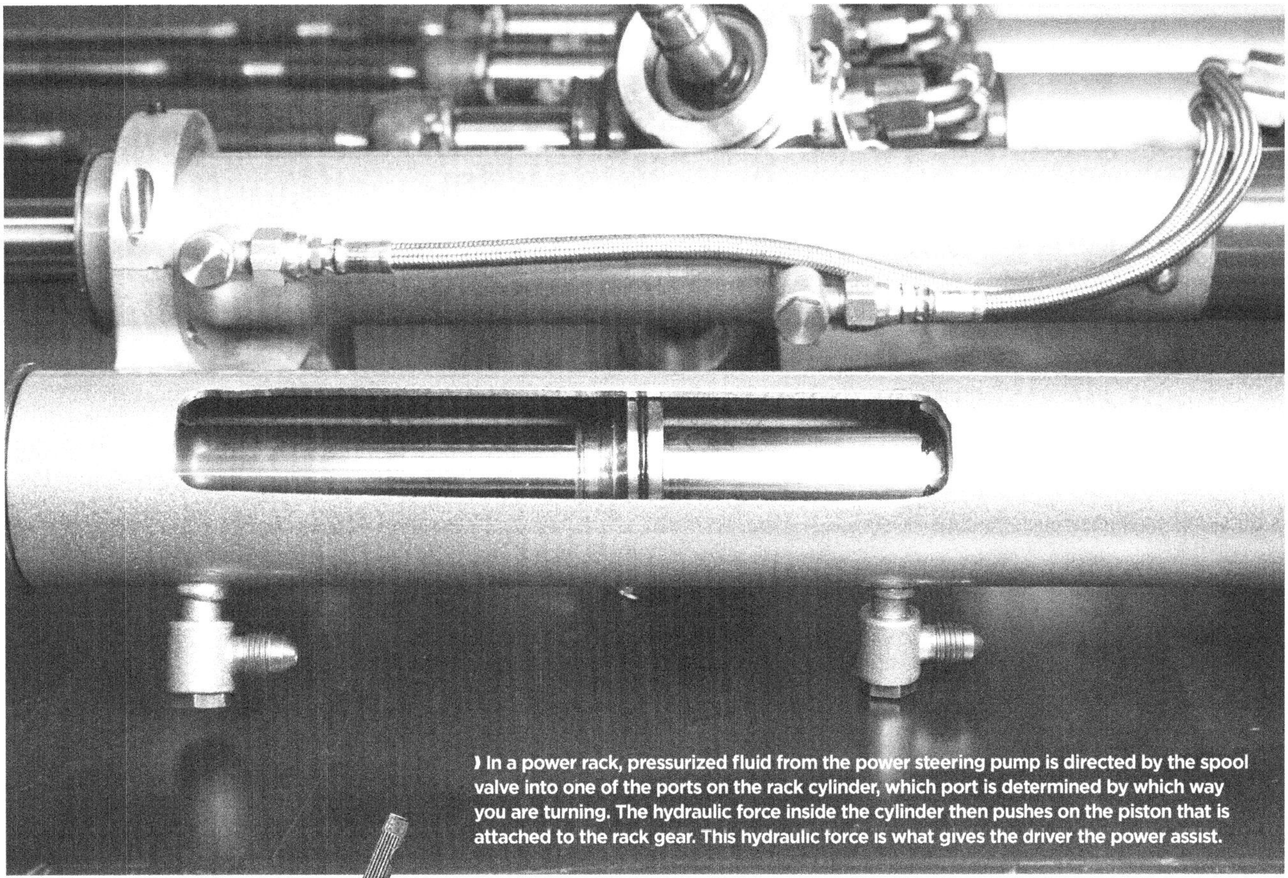

❱ In a power rack, pressurized fluid from the power steering pump is directed by the spool valve into one of the ports on the rack cylinder, which port is determined by which way you are turning. The hydraulic force inside the cylinder then pushes on the piston that is attached to the rack gear. This hydraulic force is what gives the driver the power assist.

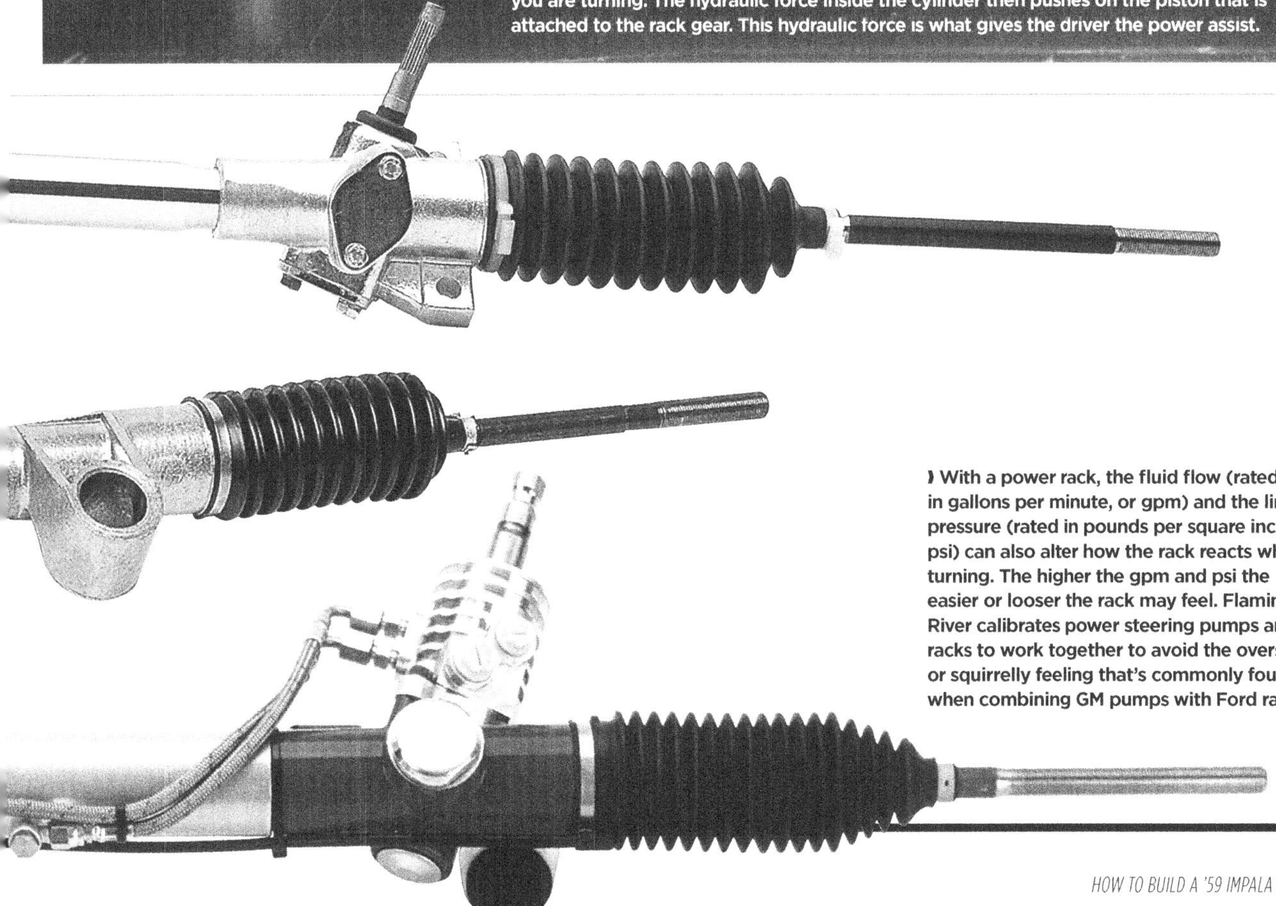

❱ With a power rack, the fluid flow (rated in gallons per minute, or gpm) and the line pressure (rated in pounds per square inch, or psi) can also alter how the rack reacts when turning. The higher the gpm and psi the easier or looser the rack may feel. Flaming River calibrates power steering pumps and racks to work together to avoid the oversteer or squirrelly feeling that's commonly found when combining GM pumps with Ford racks.

FLAMING RIVER

800 Poertner Drive
Berea, OH 44017
1-800-648-8022
www.flamingriver.com

INSTALLATION BULLETIN #2006

DESCRIPTION: FR1504P Power Shaft Kit

WARRANTY: 1 YEAR FROM DATE OF PURCHASE. REPLACEMENT OF PRODUCT DEEMED BY FACTORY TO DEFECTIVE. LABOR NOT INCLUDED.

Shaft Kit Installation

1) Measure and mark 7/8" in from each end of the DD-style shaft
2) Slide splined u-joint yoke over pinion shaft of rack and pinion.
3) Insert DD-style shaft onto opposite end of u-joint
4) Slide DD u-joint over shaft
5) Apply red threadlocker to screws
6) Insert tongue into end of steering column shaft and install through bolt. (Do not tighten the bolt at this time).
7) Adjust each u-joint to align with the 7/8" mark on the shaft. This ensures proper phasing. Be certain the set screw on the pinion-end yoke shaft is set in the groove in the pinion shaft.
8) Snug each set screw (Fig C 1-6) so it will leave a mark on to the DD shaft, to include the tongue.
9) Remove shaft kit and dimple (See Fig B) each set screw mark using a ¼" drill bit (Fig C - Set screws 1-6) ***WARNING*!** **Do not dimple the Rack and Pinion shaft; this will cause damage to your rack and pinion which IS NOT covered by warranty.**
10) Reassemble shaft kit in car. Tightening each set screw (Fig C Set screws 1-6). Use a torque wrench, tighten each screw to 25 ft-lb. Tighten jam nuts (Fig A) **NOTE:** Periodic inspection of set screws is recommended to ensure that they stay tight.

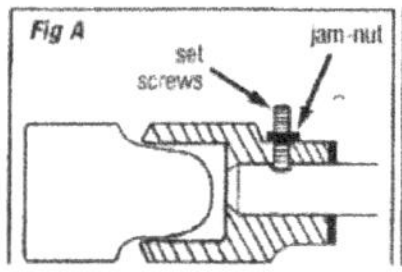

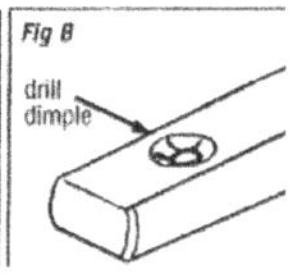

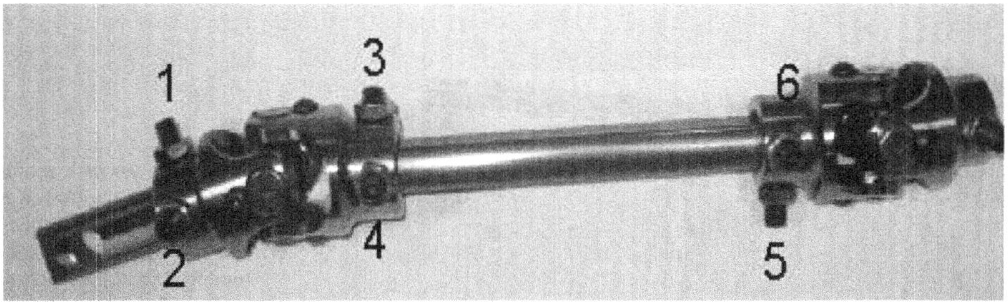

Fig C

Warranty Disclaimer

Because of their intended usage, the manufacturer makes no warranties whatsoever expressed or implied, oral or written, to purchasers of their products regarding performance, safety, fit, merchantability or length of service. Purchasers are responsible for selection of proper goods and must rely on their own skill or judgment that such goods are suitable for the purchaser's application.

Intercity Lines

Going from Point A to B with Intercity Lines

Thanks in part to the Internet, it's not unusual to buy or sell a car from one side of the country and ship it to the other. Then there are those rodders who send their cars to a distant shop for custom work, or vacationers who drive one way to an event and have the car transported the other. At one time moving a valuable car for any reason was a risky proposition, but today, it's practical, safe, and economical, thanks to Intercity Lines.

Intercity Lines started as a small, family owned business and has evolved into the nation's premier enclosed automotive transporter, thanks to the special care they give their customers' vehicles. They tell us the key to safe and secure transport of a car is proper equipment, preparation, and attention to detail.

To ensure a vehicle receives the best care possible during shipping, the pros at Intercity Lines suggest you think about the following when considering a transport company:

Reputation: It's always best to work with a known company, one with a proven track record and experience in moving special-interest vehicles. Visit websites. Get referrals. Check references. Do your homework.

Carrier versus Broker: Always work with a carrier direct. Avoid brokers. Make sure the carrier operates its own equipment—brokers don't.

Insurance: Make sure the carrier supplies a Certificate of Insurance, an agreement that states the insurance coverage as well as pick up and delivery times. Vehicles are typically insured for up to $1 million.

No Deposits: Never, never give a deposit. Legitimate carriers never ask for a deposit in advance; brokers typically do ask for a deposit. Another reason to avoid using a broker.

Tracking: Reputable companies know where their trucks—and your car—are at all times. All Intercity trucks are equipped with GPS tracking; a quick call to Intercity informs you of your vehicle's whereabouts.

Enclosed versus Open Trailer: Enclosed trailers provide maximum protection of a vehicle; it's why Intercity only uses enclosed trailers. Rain, sleet, snow, hail, sand, road debris can all damage vehicles left exposed to the elements. Moreover, cars hauled in enclosed trailers are secure against vandalism.

Liftgate versus Ramps: All Intercity trailers use hydraulic lift gates to load the vehicle into position. Vehicles with low ground clearance—exotic sports cars, lowered street rods, vintage race cars—can be damaged driving up a ramp. In addition, if a vehicle is transported often, moving up and down a ramp can strain the vehicle's driveline. Yes, clutches have been smoked on the short trip up a loading ramp.

Inoperable Vehicles: Often, cars in various stages of restoration or construction need to be shipped. No problem. Wrap spare or loose parts carefully and put them inside the car or the trunk if they will fit securely and not hamper the view from inside or risk damaging the vehicle. Extra wheels and tires can also be shipped with the car for a modest extra charge.

Winch Availablility: If a vehicle is not driveable, or if its brakes are inoperable, the transporter should use a winch to move the car on and off the truck in the safest possible manner. However, if the trailer is equipped with lift gates—as are all of Intercity's—the vehicle can be pushed onto the gate, then into the trailer.

Intercity has a fleet of tractor-trailers that routinely transport special vehicles, including the AMSOIL/*STREET RODDER* Road Tour cars.

WHEN PREPARING A CAR TO BE TRANSPORTED THE FOLLOWING STEPS SHOULD BE TAKEN:

Gas Tank: Vehicles should be transported with a quarter tank or less of gasoline in the tank. This is not only safer, it makes for a lighter vehicle. A gallon of gas weighs just under 7 pounds, so a full tank can significantly increase the overall transport weight.

Check the Antifreeze: Many vehicles based in Florida, Arizona, or Texas may be running straight water in their cooling systems. Of course, the temperature never drops below freezing in those states. However, if a car is being transported during winter to a cold-weather climate, that water could freeze in the engine block and/or radiator—with disastrous results. The solution is simple: Prior to shipment, flush the cooling system and fill it with the recommended blend of antifreeze and water.

Remove or Disconnect Battery: If not properly fastened, a battery can shift or dislodge during shipping, causing corrosive damage to wiring and sheetmetal. It could also be a fire hazard. It's also a good idea to disconnect the ground strap once the car is loaded into position.

Use the Cut-off Switch: If the car is equipped with a battery cut-off switch (many race cars have one), it should be placed to the "off" position to reduce electrical hazard. Naturally, remember to instruct the transport crew of the switch's location and operation.

Keep the Interior Free of Loose Items: The transport team needs a clear, unobstructed view from inside in order to move the vehicle safely. Exotic cars, for example, have a limited rearward view, so loading the car with personal items—that floor jack in the back seat—to save on outside shipping costs, decreases visibility even further. Moreover, such items are not covered by the transporting carrier's loss policy—only the consigned vehicle is insured against loss.

Check for Food or Perishables: Simply put, your car isn't a pantry or a fridge on wheels. Before shipping, be sure the car's interior is free of food. A Snickers can melt, a Coke can explode or puncture. Perishables can damage a vehicle.

Miller Electric Mfg. Co.

Buying Your First Welder— A Practical, Informative Guide for Do-It-Yourselfers

You may be the creative type. Or, perhaps, you have the handy knack and are just looking to tinker around … building and fixing stuff. You might have your sights set on opening a small repair shop, a studio, or new gallery. You could be a farmer or rancher, classic car enthusiast, fabricator, or woodworker … or you may just want to be able to repair or salvage something that is broken versus opting to toss it away.

Regardless of your interests, needs, or motivation, you're a do-it-yourselfer interested in buying your first welder.

Whether you already have previous welding experience, or, you're a welding novice, you are likely here because you need some help getting started. Miller Electric Manufacturing Company is dedicated to providing objective, informative, and practical information to assist you in making the appropriate choice.Relax and read on. You've come to the right place!

Getting Started

Unfortunately, there is no single welding process suitable for all applications, so let's begin with an overview of the basic processes and highlight the capabilities and advantages of each. This will help us better match a process to your specific needs.

If you have previous welding experience, feel free to skip ahead. If you are somewhat of a novice, this section will provide you with a better understanding of the types of welders available, how each performs, and the degree of welding skill required to operate each. In addition, we'll offer examples of specific applications best suited to each process.

Matching your needs and welding skills with a specific process is critical before moving on to discuss specific welder model options.

Welding Processes

The most common welding processes include Stick, MIG/Flux-cored, and TIG. Each process has its own unique set of benefits and limitations, works well in some welding applications, and not well in others. There's no "one size fits all" — and as you will soon discover — for good reason.

Stick Welding: If you learned to weld years ago, you likely learned using an arc welder. Stick welding has, for many years, been the most popular method for most home-shop welding needs. Stick welding uses an electric current flowing from a gap between the metal and the welding stick, also known as an arc-welding electrode. Stick welding is an effective method for welding most alloys or joints and can be used both indoors and outdoors, or in drafty areas. It is also the most economical welding method and largely popular because of its ability to create an effective bond on rusty or dirty metals.

Arc welding is limited, however, to metals no thinner than 18-gauge

Miller's Diversion 180 is the perfect TIG welder for the home hobbyist.

and requires frequent rod changing, emits significant spatter, and welds must be cleaned upon completion. Stick welding is also more difficult to learn and use, particularly the ability to strike and maintain an arc. Arc welders are available in either AC or DC or AC/DC—with AC being the most economical. It is used for welding thicker metals of 1/16 inch or greater. They are a good choice for farmers, hobbyists, and home maintenance chores.

MIG Welding/Gas Metal Arc Welding (GMAW): MIG welders use a wire welding electrode on a spool that is fed automatically at a constant pre-selected speed. The arc, created by an electrical current between the base metal and the wire, melts the wire and joins it together with the base, producing a high-strength weld with great appearance and little need for cleaning. MIG welding is clean, easy, and can be used on either thin or thicker plate metals.

A slight variation of MIG welding—Flux-Cored Arc Welding (FCAW): is similar in that it is also a wire-feed process but differs in that it does not require a shielding gas. This gas-free welding application uses Flux-Cored wire to shield the arc, and is a simple, efficient, and effective welding approach, especially when welding outdoors in windy conditions or on dirty materials. The process is widely used in construction because of its high welding speed and portability.

Both MIG and Flux-Cored are very easy to learn and can create extremely clean welds on steel, aluminum, and stainless. Both types have the capability to weld materials as thin as 26 gauge.

TIG Welding/Gas Tungsten Arc Welding (GTAW): TIG welding is an arc welding process that uses a non-consumable tungsten electrode to produce the weld. The weld area is protected from atmospheric contamination by a shielding gas (usually argon), and a filler metal, though some welds, known as autogenous welds, do not require it. A constant-current welding power supply produces energy that is conducted across the arc through a column of highly ionized gas and metal vapors known as plasma.

TIG welding is most commonly used to weld thin sections of alloy steel, stainless steel, and non-ferrous metals such as aluminum, magnesium, and copper alloys. The process grants the operator greater control over the weld than other welding processes, allowing for stronger, higher quality welds. TIG welding is comparatively more complex and difficult to master than other welding types, and is significantly slower.

Below summarizes each welding process. Take a quick glance at what process might be best for your general needs.

Types Of Welds

Stick Welding
- Better suited for windy, outdoor conditions
- More forgiving when welding on dirty or rusty metal
- Works well on thicker materials

MIG Welding
- Easiest process to learn
- High welding speeds possible
- Better control on thinner metals
- Cleaner welds possible with no slag to clean
- Same equipment can be used for Flux-Cored Welding

Flux-Cored Welding
- Works as well as Stick on dirty or rusty material
- Out-of-position welding
- Deep penetration for welding thick sections
- Increased metal deposition rate
- More forgiving when welding on dirty or rusty metal

TIG Welding
- Provides highest quality, precise welds
- Highly aesthetic weld beads
- Allows adjustment of heat input while welding by use of a foot control

Metal	Weld Process		
	MIG	Stick	TIG
Steel	●	●	●
Stainless Steel	●	●	●
Aluminum Alloys	●	●	
Cast Iron		●	
Chromoly			●
Copper			●
Brass			●
Exotic metals (Magnesium, Titanium, etc.)			●

Possible Projects	Average Metal Thickness
Auto body	3/16 inch or less
Trailer frames and fencing	1/4 to 5/16 inch
Farm, ranch, and landscape	5/16 to 3/8 inch
Thick structural components	Over 3/8 inch
Bicycles, lawn mowers, or tube frames	1/16 inch
Boats, cars, and motorcycles	1/16 to 1/8 inch
Hunting stands and utility trailers	1/16 to 1/8 inch
General to heavy repair	3/16 to 1/4 inch

What Process Best Fits Your Needs?

You now have a general overview of the welding processes available, but we still have a ways to go before honing in on the method that will best meet your specific needs.

Let's identify the types of welding projects and materials you will be welding the majority of the time. Will you be creating metal sculptures? Do you intend to restore an old muscle car in your garage? Does the motorcycle you bought years ago require some fabrication? Maybe you intend to do basic repair on equipment used on your farm.

Let's review the following diagram.

Taking this time up front to identify the projects that will occupy the biggest percentage of your welding activity will help you determine the specific thickness of the metal you likely weld most often, and ultimately select the most suitable welder model.

Now let's get a bit more specific. The chart above identifies which weld process you can use for each type of metal. Keep in mind that many of these materials are also processed using varying combinations of two or more metals, a process that is helpful to reinforce strength and functionality.

What Factors Should you Consider When Determining a Budget?

You likely already have an estimated budget in mind, but first-time owners should consider several important factors when deciding which welder is most appropriate.

The type of welder you purchase should be best suited for the specific functions you require as well as the projects you will be working on the most. Think about your end goal and consider opportunities to expand the usefulness of your welder. In other words, will you need more power and amperage in the future?

In addition to the cost of the welder itself, don't forget to include costs for accessories and supplies you'll need to operate your new welder. This includes welding protection (helmet, gloves, jacket, and so on) as well as gas and consumables.

One of the most versatile MIG machines available is the Millermatic 211. It features the revolutionary Auto-Set function that takes the guesswork out of selecting the proper heat range.

A Few Last Things to Keep in Mind

We've thrown a lot of information at you in a short amount of time. But now comes the fun part.

On the next page, we'll take a look at some specific welder models offered by Miller in each of the welding process categories developed with the do-it-yourselfer in mind. As you review the list of Miller offerings, take note of the varying amperage requirements of each, including power requirements and duty-cycle necessary to achieve the most effective and economical operational results.

So what is duty cycle, you ask? One way of classifying a welder's "size" is by how much amperage it can generate at a given "duty cycle." Duty cycle is the number of minutes out of a 10-minute period a welder can operate. For example, a Dynasty 350 TIG unit can deliver 300 amps of welding output at a 60 percent duty cycle. It can weld continuously at 300 amps for six minutes, and then must cool down during the remaining four minutes to prevent overheating.

To see if a machine meets your do-it-yourselfer needs, consider the following power classifications: Light Industrial products typically have a 20 percent duty cycle and a rated output of 230 amps or lower. More industrial products typically have a 40 to 60 percent duty cycle and a rated output of 300 amps or lower.

The Miller lineup of welders also incorporates several exclusive features, including their Adaptive Hot Start technology, which means you no longer have to worry about those difficult-to-strike stick electrodes. The Multi Voltage Plug (MVP) allows you to connect to common 115V or 230V power receptacles without the use of any tools. Simply choose the plug that fits the receptacle and connect it to the power cord.

Don't feel you have to make a purchasing decision right now. Take some time to define your needs. If you have questions about welding processes, benefits, limitations, or operation, or something is unclear, call Miller. And when you're ready to match a specific model with the task, hobby, or business they will suggest the model or product that is the best for you.

Let's Go Shopping:
Stick Welders

To review, the Stick welding process is ideal for general construction, maintenance and repair, shipboard installation/repair, farm and ranch applications, home repair, and plant fabrication/repair. Miller compact designs allow for greater portability. When considering duty cycle for a Stick machine, note that most Stick electrodes are consumed in less than two minutes. About 80 percent of all Stick welding is done with an 1/8-inch-diameter electrode, which takes about one minute to consume.

Suggested Products:
- Maxstar 150S
- Thunderbolt XL 225/150 AC/DC

MIG/Flux-Cored Welders

Again, the MIG welding process is ideal for general maintenance and repair, farm and ranch applications, home repair, and autobody. Key features exclusive

Plasma cutters have made cutting torches virtually obsolete. Miller offers a variety of machines ranging in applications from home shops to production facilities

Something that makes welding much easier is the self-darkening helmet—Miller offers a variety of them.

to the Millermatic family of MIG welders include:

Auto-Set: A breakthrough control that automatically sets your welder to the proper parameters. Auto-Set offers all-in-one MIG minus the guesswork. Simply set the wire diameter and material thickness and you're ready to start welding.

Standard Built-In Solid-State Contactor Circuit: Makes wire electrically "cold" until trigger is pulled. This makes the unit safer and torch easier to position before starting to weld.

Thermal Overload Protection: Shuts down the unit and activates over-temperature light if airflow is blocked or duty cycle is exceeded. Automatically resets when fault is corrected and unit cools.

Suggested Products:
- Millermatic 140 Auto-Set
- Millermatic 180 Auto-Set
- Millermatic 211 Auto-Set with MVP
- Millermatic Passport Plus
- Millermatic 212 Auto-Set
- Millermatic 252

TIG Welders

Miller Diversion AC/DC TIG welders contain all the features you need to make welding easy, fun, and affordable. The TIG welding process is ideal for home repair, garage and shop jobs, autobody, chassis/frame fabrication, aluminum oil pans, stainless exhaust, metal art, and sheetmetal applications.

Key features exclusive to Miller include:

- Easy-to-Understand Operator Interface
- It's as easy as 1, 2, 3 to set your machine and start welding!
  1. Power up
  2. Select material type
  3. Set material thickness range

Inverter Technology

By utilizing an inverter-based AC/DC power source, a more efficient (consumes less energy) and consistent welding arc is provided than older transformer-based machines.

Eliminates Waste of Expensive Shielding Gas

Miller's Auto-Postflow feature optimizes postflow time based on welding amperage. This allows for proper shielding of both the weld end zone and electrode every time without waste of gas or need for adjustment.

Reduced Noise, Maintenance, and Operating Costs

Miller's Fan-On-Demand cooling system operates only when needed. This cuts down on fan noise, along with reducing power consumption and contaminants drawn into the machine.

Suggested Products:
- Diversion 165
- Diversion 180
- Dynasty 200 Series

To access a thorough listing of all Miller welder models, supplies, and related accessories, visit millerwelds.com/products. You may also request a free full-line catalog of welding and cutting equipment by visiting the literature request page at millerwelds.com. Here you can also find information on Miller's welding protection and safety apparel, safety guidelines and resources, distributor network, product registration, and other welding resources.

Speedway Motors

Founded by the late "Speedy" Bill Smith, Speedway Motors is said to be the oldest speed shop in America. In the beginning Speedway Motors was housed in a tiny building on O Street in Lincoln, Nebraska—today it occupies a 520,000, 46-acre facility that also includes the Smith Collection Museum of American Speed. This miraculous museum displays one-of-a-kind race cars, engines, equipment and parts.

Speedway Motors' huge inventory and ability to ship quickly made it our choice for the "go-to" place for many of the parts need to get our Road Tour car on the road. ★

★14

Lokar

They've Been With Us From The Beginning

Bundles of various sizes and types of aluminum will soon be transformed into parts for Lokar's array of products.

One of the best-known and most respected names in the Street Rod industry is Lokar. Founded in 1988, their motto has always been "Quality ... Plain & Simple." From shifters to throttle pedals and everything in-between, Lokar products are designed and manufactured in-house by street rodders for street rodders.

Since the inception of the Road Tour program Lokar products have been found on every *STREET RODDER* Road Tour car, which is not only a testimony to their support of the hobby but the quality of their USA-made products as well; you won't find an inferior off-shore product wearing their logo.

Our Road Tour cars are built to the highest standards of fit and finish, as are all Lokar products, so they always look like they belong. But there's more to the Road Tour than looking good; our cars accumulate 25,000 miles or more in a summer's worth of cross-country, border-to-border flogging, so we've tested Lokar's products repeatedly and they've always passed with flying colors. Like they say, "Quality ... Plain & Simple." ★

› Lokar's state-of-the-art manufacturing facility is spotless and well organized.

› Those CNC machines in the previous photo stay busy cranking out quality parts like these connections used on transmission dipstick tubes.

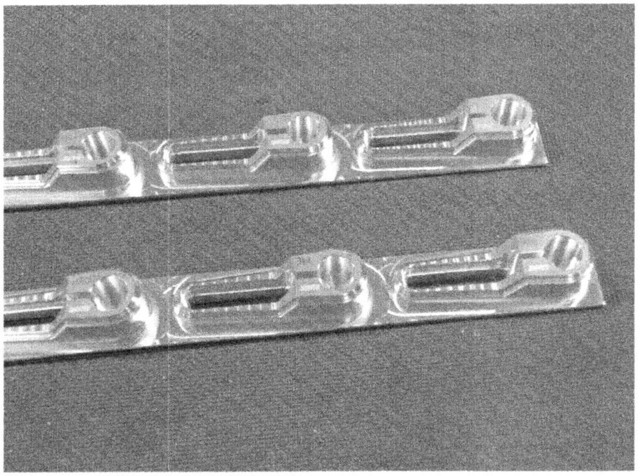

› Here are partially machined arms for automatic transmission shifters.

› The Eliminator floor-mount throttle pedals eliminate routing cables through by using an under-the-floor cable mounting bracket.

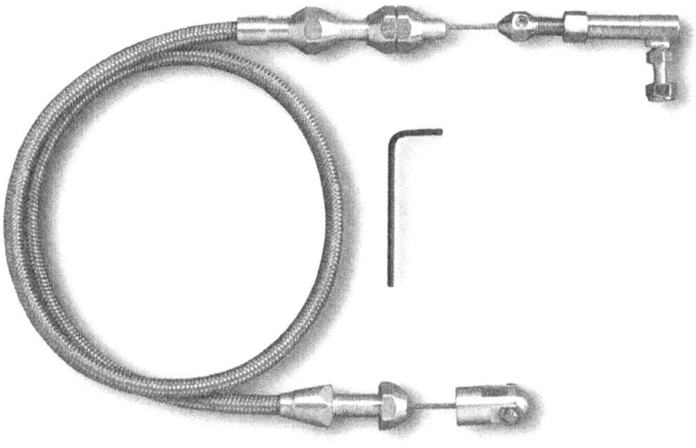

› Lokar Throttle Cables are available in either stainless steel braided housing or black housing. They are designed with a specially extruded liner for the extended life of the cable.

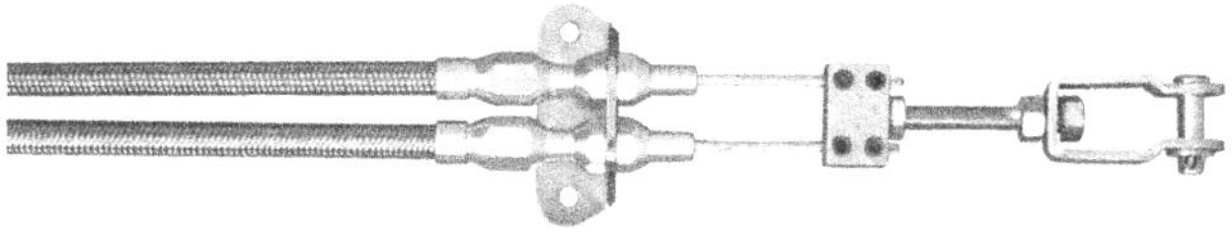

› Shown here are cables designed for use with Lokar's Floor Mount Emergency Hand Brake. The kit comes equipped with quality one-piece aluminum adjusters and aluminum ferrule to eliminate fraying of stainless housing. Outer housing is Teflon-lined for smooth operation and stainless inner cable resists moisture.

CHAPTER FIFTEEN

2014 AMSOIL/*STREET RODDER* Road Tour '59 Chevrolet Impala

Sitting on 20,000 miles, it's amazing how quickly the miles add up

By Brian Brennan | Photography by Randy Lorentzen

HotRods
BY DEAN
PHOENIX ARIZONA

Starting its 19th year, the AMSOIL/*STREET RODDER* Road Tour embarks on its most ambitious build yet. The beginnings of this year's '59 Chevy Impala were found in the deserts of Arizona and rescued by Dean Livermore of Hot Rods by Dean (HRBD) who would continue the tradition by handling the build. Before the build we looked to Eric Black to give us a peek into the future of what our '59 could be.

HRBD was in charge of the build, which included any and all added fabrication, body- and paintwork, electrical, and final assembly. The upholstery was handled down the street by Glenn Kramer at Hot Rod Interiors by Glenn (they handle many of the HRBD projects). All of the efforts by HRBD were aided by Eastwood Company tools and Miller Electric Manufacturing Company welding equipment, benches, and safety equipment. Speedway Motors also supplied all of those last-minute detail items, such as the battery, starter, valve covers, steering wheel, a draw full of miscellaneous fittings, and related hardware. It should also be mentioned that many of the components we used in this build are readily available from Speedway Motor's mail-order catalog from their online website.

Armed with a trailer load of sheetmetal and visions of what could be, it was off to the Roadster Shop via Intercity Lines to have our chassis built. Starting with an RS Fast Track chassis outfitted with both IFS and IRS (RS uses Strange Engineering third members). The FT also employs RS by Penske adjustable coilover shocks, Wilwood disc brakes, and a Flaming River rack-and-pinion steering. Resting within the Roadster Shop 'rails is a Shaver Racing Engines LS motor outfitted with a COMP Cams camshaft and valvetrain along with an Inglese LS 8-Stack induction held in tune by a FAST computer linked to a Hughes Performance transmission 4L80E, which in turn twists a Dynotech driveshaft making up the powertrain. As with any hot rod it takes lots of hardware to hold one of these together and a '59 Chevy requires more than most. We turned to Automotive Racing Products (ARP) for the stainless 12-point hardware that included nuts and flat and lock washers throughout.

STREET
RODDER
AMSOIL

ROAD TOUR
2014

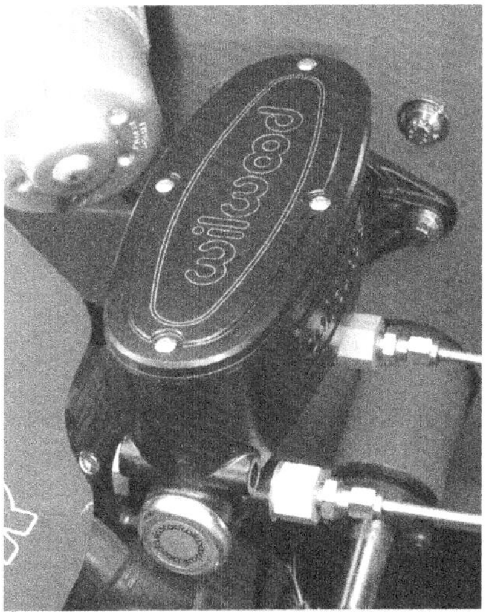
wilwood

There are lots of focal points on our AMSOIL/*STREET RODDER* '59 Chevy Road Tour car but the wheel and tire combination has most assuredly grabbed everyone's attention. Circle Racing Wheels supplied the 19s and 20s that measure 10 inches wide in front and 12 inches in back. To this Michelin Pilot Super Sports currently found on Dodge Vipers is the rubber mounted, measuring 285/35R19 in front and 345/30R20 in back. The wheels are 5-on-4.75 pattern with 5.25-inch backspacing in front and 6-inch backspacing in the rear.

Next up on the hit parade of eye-grabbing points of note is the PPG Ferrari Red (color: DBC908276W; white sealer: NCS2003); and we do mean Ferrari Red. Hot Rods by Dean body shop staff of Sean Rosic handled the prep and bodywork while Frank Gracia applied the color. Other HRBD paint projects include the sealing of the interior before plenty of Dynamat was applied and then undercoating on the bottom exterior utilizing Eastwood Company sealer.

Beginning with an RHS block, the reciprocating pieces are made up of a combination of SCAT crank and rods and COMP Cams camshaft and valvetrain, all held together with a bucket full of ARP hardware—inside and out.

Next up on the eye candy parade is the V-8. And what an LS engine it is, sporting 427 inches and producing 622 hp and 620 lb-ft of torque from the Shaver Racing Engines. Making sure all of the moving parts stay well lubricated we looked to our sponsor AMSOIL for engine oil and filter who also supplied the radiator coolant, power steering fluid, differential lube, and transmission fluid. Beginning with an RHS block, the reciprocating pieces are made up of a combination of SCAT crank and rods and COMP Cams camshaft and valvetrain, all held together with a bucket full of ARP hardware—inside and out. An Inglese 8-Stack induction system is used and kept in tune by the FAST EZ-EFI 2.0 system. Supplying

Text continued on page 171.

POWER
STEERING
USE ONLY
APPROVED

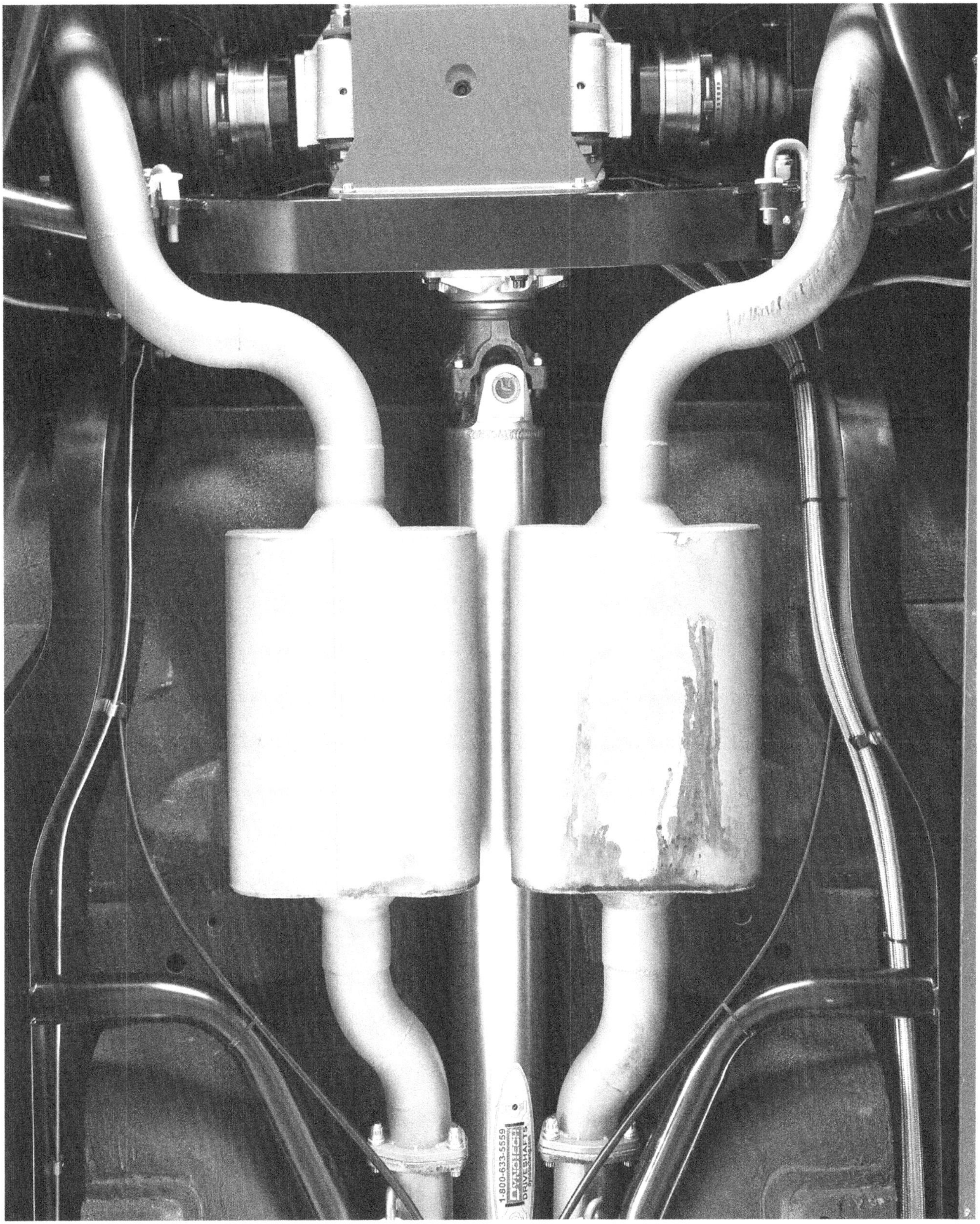
1-800-633-5559
DRIVESHAFTS

Text continued from page 166.

the Inglese-equipped Shaver LS is a full complement of Aeromotive Phantom fuel pump, filter, lines, and fittings that begin at the Rock Valley tank, which is another item that RS provides. Removing the spent gases from the LS motor falls to a combination of Flowmaster headers, exhaust pipes, and mufflers that all received a liberal covering of Jet-Hot thermal coatings, ensuring good looking and long-lasting components. However, there is one external sound that any hot rodder will enjoy and those come from the exhaust "tunes," and this LS plays a mighty fine chorus through the Jet-Hot–coated Flowmaster headers, stainless steel pipes, and mufflers.

It should be noted that the engine must maintain its cool too! U.S. Radiator supplied the copper and brass radiator, complete with dual electric fans and shroud. Providing double cooling—partly for the engine and partly for the passengers is the Vintage Air Front Runner. This system runs the water pump and air conditioning compressor along with the power steering and alternator.

Painless Performance Products was called upon to supply the required electrics to operate the Dakota Digital custom face gauges, Custom Autosound stereo, complete with iPod hookup, Specialty Power Window electric windows, and windshield wiper system. From here a Flaming River tilt steering column and a Lokar floor shift that pops up through a center console from Classic Industries, who also supplied much of the interior and exterior trim, can also be seen. Of course many trim items were retained from the original car and these pieces were lovingly restored through polishing or freshly chrome-plated steel by Sherm's Custom Plating.

Text continued on page 175.

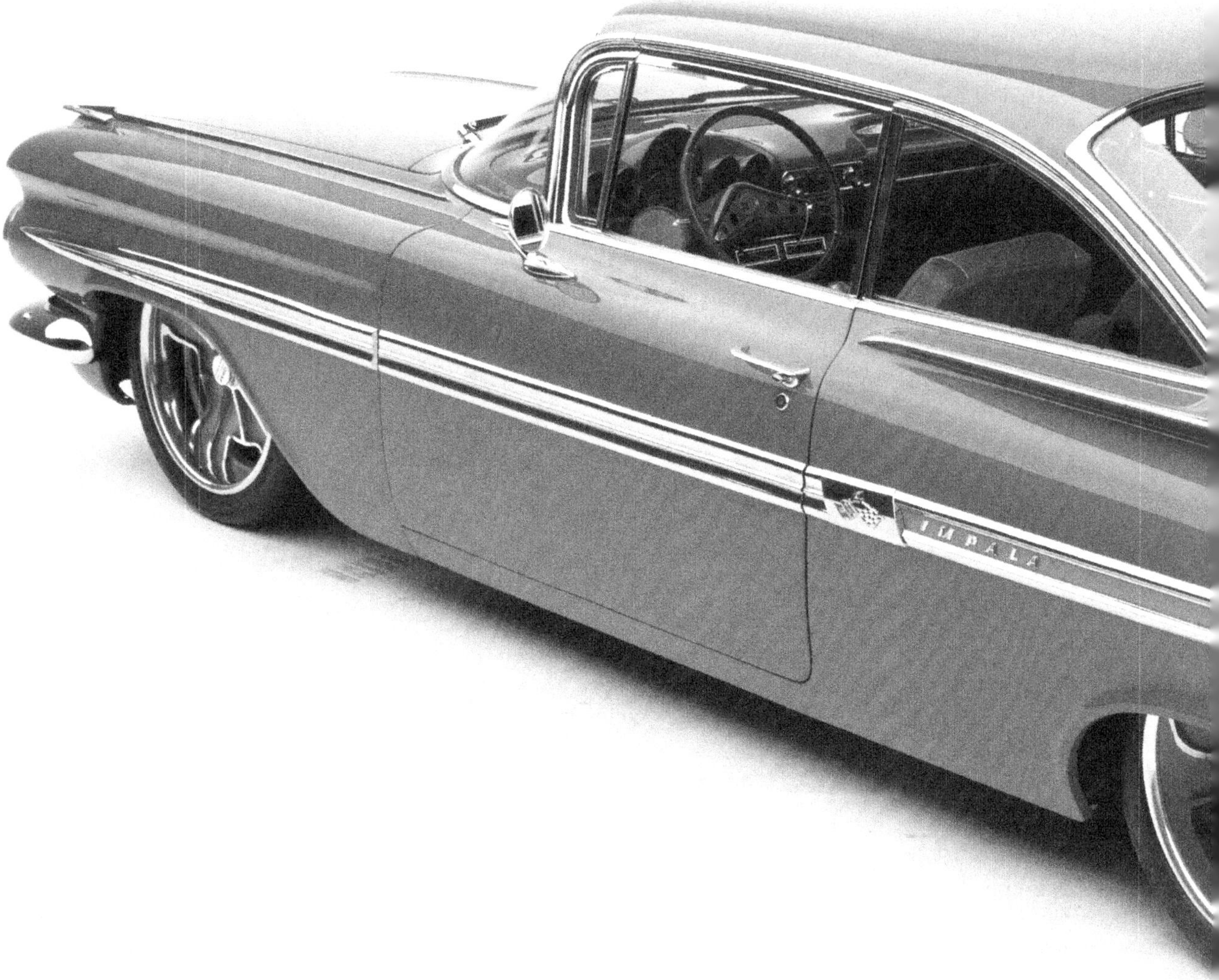

At the time of printing the "heavy Chevy" was closing in on 20,000 miles of highway cruising and hanging in there at 17 mpg. Not bad for a monster of a car with 622 hp.

BAND
MODE
8
10
12
14
MENU
M1/D-UP
M2/D-DN
M3/T-UP

Text continued from page 171.

This included much of the factory exterior and interior trim pieces, as well as the front and rear bumpers. The headlights, buckets, trim, and all of the grille components are from Classic Industries.

The interior is custom indeed and while featuring a number of original items it becomes apparent that Hot Rod Interiors by Glenn paid special attention to the Impala, not the least of which are the custom-upholstered Procar (division of SCAT) front buckets and the modified rear bench seat. Something you didn't see on a '59 were buckets with a center console and with a little "massaging" by HRBD and Hot Rod Interiors by Glenn they were able to adapt center console from a '63 Chevy Impala SS provided by Classic Industries (CI). To the newly fit CI console you will also see a Lokar Performance Products shifter. More Lokar goodies include the blacked-out throttle and brake pedals and e-brake assembly. The CI door panels received more personal attention from Kramer in the way of later model Chevy armrests, again from CI, and custom trim application.

An interior is only as comfortable as its own "climate." Making sure the temps are cool when they should be and toasty when called for, a Vintage Air A/C and heat system were installed, while resting beneath the red carpeting is plenty of Dynamat heat and sound insulation. Furthering the comfy surrounds, the full-size Impala sports lots of tinted glass from CI while SoffSeal provided the definitive seal against any intrusion from the outside.

Well, that's 19 year's worth of Road Tour cars and the 2014 AMSOIL/*STREET RODDER* Road Tour '59 Chevy Impala has proven to be one of the more popular cars, and assuredly the longest and heaviest! We should mention, at the time of printing the "heavy Chevy" was closing in on 20,000 miles of highway cruising and hanging in there at 17 mpg. Not bad for a monster of a car with 622 hp. *

2014 AMSOIL/ STREET RODDER ROAD TOUR SPONSORS

To produce a vehicle like the AMSOIL/*STREET RODDER* Road Tour Chevrolet it takes the finest craftsmen and the highest quality products—and we certainly had both. Here are those that deserve the credit for our '59 Impala.

Aeromotive Inc.
Fuel System
(913) 647-7300
aeromotiveinc.com

AMSOIL INC.
Lubricants & Filtration/Title Sponsor
(715) 392-7101
amsoil.com

Automotive Racing Products
Fasteners/Presenting Sponsor
(805) 339-2200
arp-bolts.com

Circle Racing Wheels
Billet Aluminum Wheels
(559) 251-1127
circleracing.com

Classic Industries
Restoration Parts Supplier
(888) 209-0773
classicindustries.com

COMP Cams
Camshafts/Valvetrain
(800) 999-0853
compcams.com

CPG/FAST
Engine Management
(877) 334-8355
fuelairspark.com

Custom Autosound
Audio Electronics
(800) 888-8637
customautosound.com

Dakota Digital
Gauges
(800) 852-3228
dakotadigital.com

Dynamat
Insulation & Noise Control
(513) 860-5094
dynamat.com

Dynotech Driveshafts
Driveshaft
(800) 633-5559
dynotecheng.com

Eastwood Company
Tools
(800) 343-9353
eastwood.com

Flaming River
Steering Components
(866) 815-3669
flamingriver.com

Flowmaster
Exhaust & Headers
(707) 544-4761
flowmastermufflers.com

Hot Rod Interiors by Glenn
Interior
(623) 872-2242
hotrodinteriorsbyglenn.com

Hot Rods by Dean
Builder
(800) 362-9709
hotrodsbydean.com

Hughes Performance
Transmission
(800) 274-RACE
hughesperformance.com

Intercity Lines
Transport
(800) 221-3936
intercitylines.com

Jet-Hot
Thermal Coatings
(800) 432-3379
jet-hot.com

Lokar Inc.
Shifter & Accessories
(865) 966-2269
lokar.com

Miller Electric Mfg. Co.
Welding & Cutting Products
(800) 426-4553
millerwelds.com

Painless Performance
Wiring & Electrical System
(800) 423-9696
painlessperformance.com

PPG Industries
Paint & Coatings
(440) 572-2800
ppgrefinish.com

Procar by Scat Enterprises
Seats
(310) 370-5501
procarbyscat.com

Roadster Shop, The
Chassis, Front Suspension & Rearend
(847) 949-7637
roadstershop.com

RS by Penske
Shocks
(847) 949-7637
roadstershop.com

Scat Enterprises
Crankshaft/Rods
(310) 370-5501
scatenterprises.com

Shaver Racing Engines
Engine
(310) 370-6941
shaverspecialty.com

Sherm's Custom Plating
Chrome Plating
(916) 646-0160
shermsplating.com

Soffseal Inc.
Weatherstrip
(513) 367-0028
soffseal.com

Specialty Power Windows
Power Windows & Wipers
(800) 634-9801
spwkits.com

Speedway Motors
Speed Shop
(855) 682-9226
speedwaymotors.com

U.S. Radiator
Radiator, Shroud & Electric Fans
(323) 826-0965
usradiator.com

Vintage Air Inc.
Air Conditioning
(800) 862-6658
vintageair.com

Wilwood Engineering
Disc Brakes
(805) 388-1188
wilwood.com

Made in the USA
Monee, IL
06 November 2026

67383143R00098